MECHANICAL SUTURES IN OPERATIONS ON THE SMALL & LARGE INTESTINE & RECTUM

MECHANICAL SUTURES
IN OPERATIONS ON THE
SMALL & LARGE
INTESTINE & RECTUM

Felicien M. Steichen, M.D., F.A.C.S.
Professor of Surgery
New York Medical College
Valhalla, New York

Ruth A. Wolsch, R.N., B.S.

Published by: Ciné-Med, Inc.
                     127 Main Street North
                     Woodbury, CT 06798

ISBN: 0-9749358-2-4

Medical Illustrators:
    William Baker
    Robin Lazarus
    Virginia Ferrante

TABLE OF CONTENTS

MECHANICAL SUTURES IN OPERATIONS ON THE SMALL AND LARGE INTESTINE AND RECTUM

Sutures, manual or mechanical, are an essential component of the reconstructive phase in a surgical procedure. They should therefore possess technical versatility, provide support for the surgeon's skills in dealing with normal anatomy and pathologic changes, contribute to reliable wound healing and present no intrinsic short- or long-term liabilities. In the current state of the art, the types and variety of materials available to suture human tissues are of comparable merit; with exceptions such as the demonstrated advantage of staplers in confined anatomical spaces and the clear indication for specific materials in vascular surgery.

In operations on the small and large bowel as well as the rectum, the cost advantage of manual sutures is offset by reduced procedure time, technical reliability, and reproducible high-quality results achieved with stapling. Human imperfections are eliminated, especially in complex anatomical repairs, as the demanding curative dissection and excision blend into the equally exacting repair of form and function. Additionally, mechanical sutures facilitate reconstructive procedures in areas that are difficult to access in colorectal surgery. As an example, the low anterior resection and anastomosis with the circular stapler in patients with adequate tumor margins has safely replaced the abdominoperineal amputation performed to excise low rectal cancers. At this critical level, a variety of stapling techniques provide reliable bowel continuity without the need for a permanent colostomy, and shelter the patient from the higher incidence of debilitating and potentially lethal anastomotic leaks associated with a technically difficult manual anastomosis.

While the wound closure and anastomotic techniques and materials should be sufficiently flexible to adapt to a variety of repair methods and substitute organ creations dictated by the intraoperative circumstances of the therapeutic first stage, the opposite is not acceptable in any given operation: namely the subordination of this first curative stage to the available reconstruction potential. This means that the surgeon has to be equally skilled in manual and mechanical suture techniques.

Surgical principles and safeguards must also be respected, regardless of the proposed or planned repair. These include: bowel preparation and peri-operative antibiotic prophylaxis, non-invasive and invasive monitoring of vital functions as dictated by the patient's age and co-morbidity factors, meticulous dissection and no-touch technique as indicated and feasible, radical primary and lymph node excision in cancers guided by pre- and intraoperative staging with endo-sonography, total mesorectal excision (TME) in rectal cancer, attention to fluid, electrolyte and blood loss replacement, and creation of ostomies if indicated.

Reconstructive techniques that are not compatible with these requirements should be eliminated from the surgical/operative repertoire. While the instruments used to introduce mechanical sutures appear cumbersome, in fact they actually facilitate eye-hand coordination by placing the surgeon's hands at the very periphery and allowing an unobstructed view of the required action.

MECHANICAL SUTURES IN SMALL AND LARGE BOWEL OPERATIONS BY LAPAROTOMY

In the mid- to late-1960s, when American designed and manufactured instruments became available, the linear stapling and cutting instrument for side-to-side anastomosis or simultaneous closure and transection of bowel lumina had the appearance of a most ingenious instrument in search of an application that would justify the research and development effort involved in its creation. The NZhKA, its much simpler Russian precursor, had obviously served for true side-to-side anastomosis, such as in gastroenterostomy with the stomach in continuity or after distal gastric resection with Billroth II reconstruction and in enteroenterostomy for bypass of non-resectable bowel tumors.

However, in the age of retreat from subtotal gastrectomy for the treatment of duodenal ulcers and the replacement by various, increasingly nerve-sparing, forms of vagotomy and pyloroplasty, if indicated, as the preferred drainage procedure, a side-to-side anastomosing stapler assumed the semblance of a fancy gadget rather than a useful, progressive instrument. Added to this evolution in the operative treatment of duodenal ulcer and its ultimate replacement by pills and potions, was the fortunate decrease in the incidence of gastric cancer in the Western world, reducing the need for radical subtotal and total gastrectomy and reconstruction of digestive tract continuity with end-to-side or side-to-side anastomoses. Finally, the preference for palliative excision rather than bypass in incurable bowel cancers pointed to the need for mechanical anastomotic techniques that would eliminate any potential for the formation of a blind pouch, a long-term complication of the classic side-to-side anastomosis. At the time, the circular end-to-end anastomosing instrument was not even on the drawing board, and its Russian versions, the PKS, SPTU, and KS, placing a single circle of staples, were unreliable and always required a reinforcing layer of manual sutures, which prohibited their use deep in the pelvis.

Yet the quality and beauty of anastomoses obtained with the GIA instrument made its use in small and large bowel surgery very desirable, if a blind pouch could be avoided. Considering the GIA instrument to be the modern equivalent of the clamp (enterotome) employed by Dupuytren and von Mikulicz (early and late 19th century) to re-establish bowel continuity by crushing the spur formed by the merged afferent and efferent bowel walls of a double barrel ostomy, it appeared to the senior author that a primary side-to-side anastomosis could be created, functioning like an end-to-end anastomosis, without the risk of a blind pouch (1968). This would also facilitate the joining of mismatched bowel ends, unlike a true end-to-end anastomosis where uneven bowel calibers can create a challenge to the surgeon's dexterity.

After bowel resection has been accomplished, the GIA instrument makes possible the telescoping of the various stages of the 19th century procedures into one step. It serves for the serosa-to-serosa anastomosis and division of the double spur, created by the apposition of the bowel stumps, left after resection. The resulting common lumen is then closed transversely in an everting, mucosa-to-mucosa fashion with a linear stapler (1968). The somewhat serendipitous discovery of a new usefulness for the GIA instrument, with the functional end-to-end anastomosis in small and large bowel operations, opened the clinical function of mechanical sutures to a series of new concepts in surgical techniques that had been previously explored in the experimental laboratory and found to be valid and safe, e.g., the reliable healing of everting, mucosa-to-mucosa closure of hollow viscera and the crossing and overlapping of staple lines without causing necrosis; both features used to close the GIA introduction site transversely with a linear stapler. While everting manual sutures had been known to heal since the work of Travers in 1812, crossing of suture lines was a new concept made possible by the small vessel sparing, B-formation of staples.

Additionally, this new mode of joining bowel ends stimulated a cottage research effort, especially into the geometry of the functional end-to-end anastomosis that would be best suited to a variety of sometimes unexpected intraoperative circumstances. As a result of this research, Turbelin and Welter demonstrated (1980) that the cross-section of this anastomosis could be increased almost three-fold, depending on whether the GIA exit gap was closed in an oval (Figure V-2,D,E) or V-shape (Figure V-1,C,D,E).

Earlier, Ravitch (1974) had already explored a different method in the construction of this anastomosis, by reversing the sequence of resection followed by reconstruction, adhered to in most operations. In this modification of the original technique, the separation of the fully dissected specimen from the healthy remaining bowel was incorporated into the closing stage of the functional end-to-end anastomosis. The anastomosis was performed, -first,- into the viable afferent and efferent bowel loops, proximal and distal to the specimen, kept temporarily in continuity, and used as a convenient handle to accomplish the anastomosis. The transverse linear closure of the GIA introduction gap and subsequent transection of both afferent and efferent specimen loops, peripheral to the linear stapler used as a guide for this transection, present the surgeon with a wide selection of anastomotic cross-sections, depending on the choice between a narrow or wide V-shaped, functional end-to-end anastomosis. This technical modification, called "anastomose-resection integree" (anastomosis first, - resection second) by Welter (1981) who had developed a similar technique independently and concurrently for the gastroenterostomy in a Billroth II gastric resection, was to gain great popularity at all levels of the GI tract and later became a favorite way of establishing continuity in laparoscopically assisted colon resections.

Though the use of mechanical sutures can enhance and, in some cases, allow specific surgical procedures, adherence to optimal surgical technique and awareness of precautions in the use of stapling instrumentation should be employed. With use of the linear, linear-cutting, or circular stapling instruments, the tissue should be adequately prepared to prevent incorporation of extraneous tissue while allowing adequate nutrition to the cut edge. The staple size should be chosen dependent on the thickness of the tissue to be closed and should compress the tissue sufficiently to provide hemostasis without crushing.

ANATOMICAL SIDE-TO-SIDE AND FUNCTIONAL END-TO-END ANASTOMOSIS INTO OPEN BOWEL LUMINA

This technique is well suited to small bowel anastomoses, where the caliber of the bowel is proportionate to the thickness of the arms of the linear anastomosing instrument, and the short duration of exposure to the open small bowel does not represent a source of contamination in the average patient. The technique is especially helpful in the child with compatible bowel lumina, where a manual anastomosis would prolong the exposure to open bowel and the smallest end-to-end anastomosing instrument would be incompatible with the bowel caliber. It is also of particular advantage in cases with discrepancies in intestinal caliber, e.g., small to large bowel anastomoses.

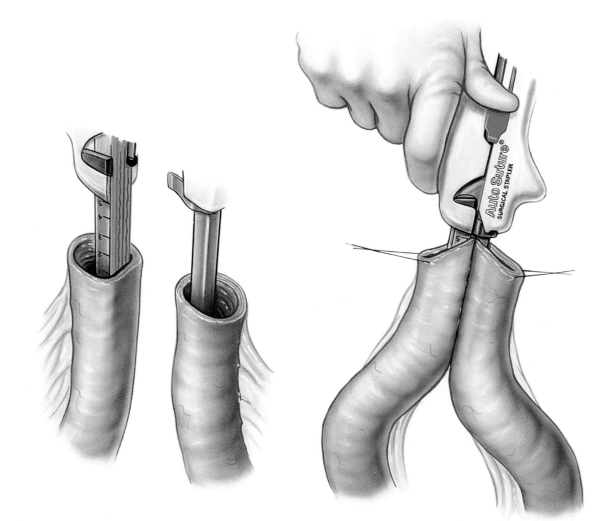

Figure V-1,A: After resection of the specimen, the remaining bowel loops are arranged with their antimesenteric borders facing each other. As in all side-to-side anastomoses, the vessel bearing mesentery is left intact, right up to the level of bowel transection, and its connection to the bowel is kept opposite to the proposed anastomosis. The forks of the linear anastomosing instrument are then placed, one into each bowel lumen, with their working surfaces against the inside of the antimesenteric bowel wall. During this step, "wandering" of the instrument should be avoided so as to prevent injury to the mesentery and vessels at the tip of the anastomosis because the bowel was not kept in proper parallel alignment, with antimesenteric surfaces apposing each other.

Figure V-1,B: The instrument halves are mated, bringing the evenly aligned antimesenteric bowel walls in apposition, and the knife-staple forming assembly is activated, creating a one step side-to-side anastomosis.

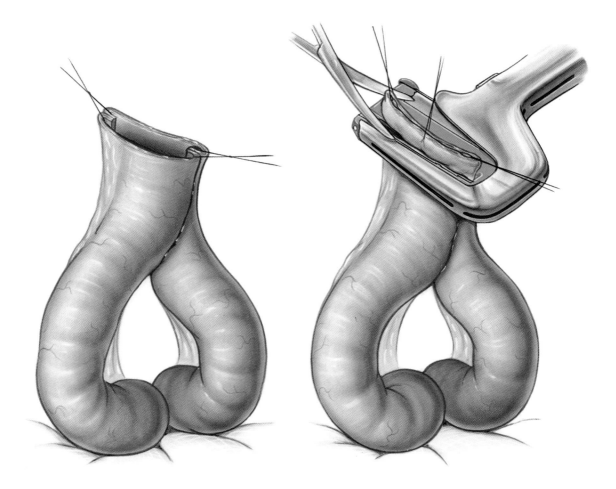

Figure V-1,C: Following inspection of the anastomosis for hemostasis, the staple lines at the opened end of the anastomosis are separated and maintained by stay sutures or Allis clamps in a wide V configuration.

Figure V-1,D: The open end of the anastomosis is closed transversely with a linear stapler, ensuring that all layers of both bowel ends and the ends of the GIA staple lines are included in this closure. The excess tissue is excised along the instrument edge prior to removing the instrument. An anchor suture may be placed at the distal end of the anastomosis for additional security.

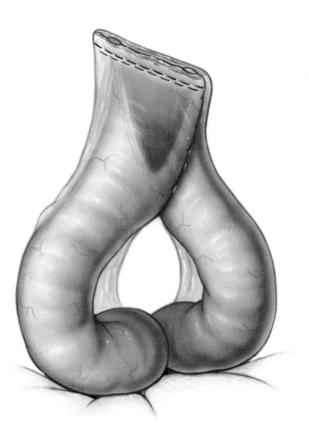

Figure V-1,E: The finished anastomosis is in the wide V mode, providing a cross-section that precludes stricture formation, which is a distinct advantage in narrow small bowel.

ANATOMICAL SIDE-TO-SIDE AND FUNCTIONAL END-TO-END ANASTOMOSIS INTO CLOSED BOWEL LUMINA

This technique represents our preferred approach to large bowel anastomoses. The closure of all bowel lumina, while the dissection and excision of the specimen are in progress, protects the peritoneal cavity from contamination, especially in emergency operations where the pre- and intraoperative bowel preparation is far from ideal. Furthermore, the narrow access into the bowel lumina, obtained by excising the antimesenteric corners of the preliminary bowel closures, guides the instrument fork along the inner aspect of the bowel wall, opposite the mesentery, and adds a safety feature to the placement of the GIA instrument.

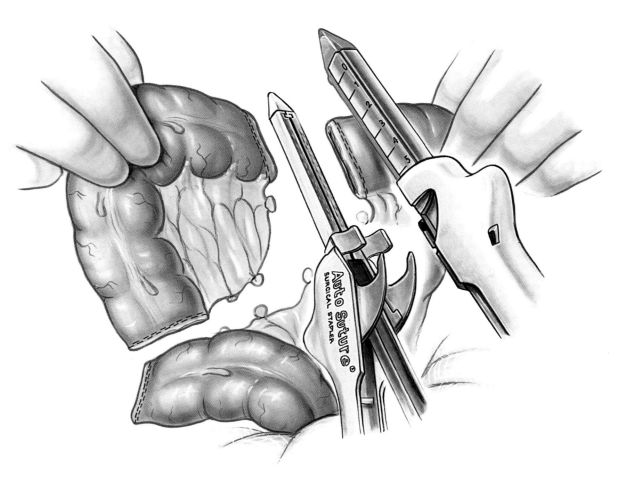

Figure V-2,A: The specimen is resected between two applications of the linear anastomosing instrument, which simultaneously places two staggered rows of staples on each side of the transection, to close both the specimen and the remaining bowel ends.

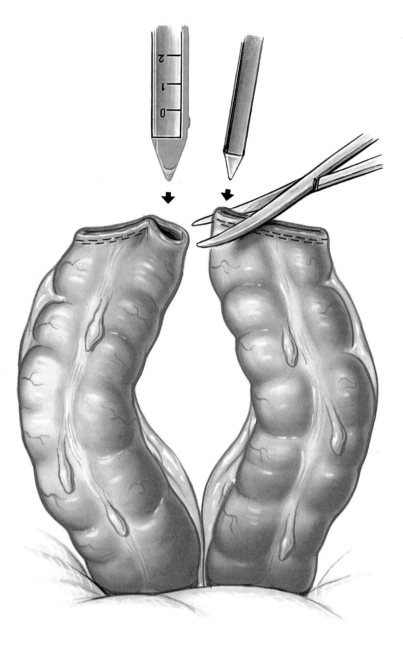

Figure V-2,B: The antimesenteric corners of the in situ closures are excised and the forks of the GIA instrument are inserted, one into each lumen. To ensure maximal stomal size, the forks should be fully inserted into the lumina.

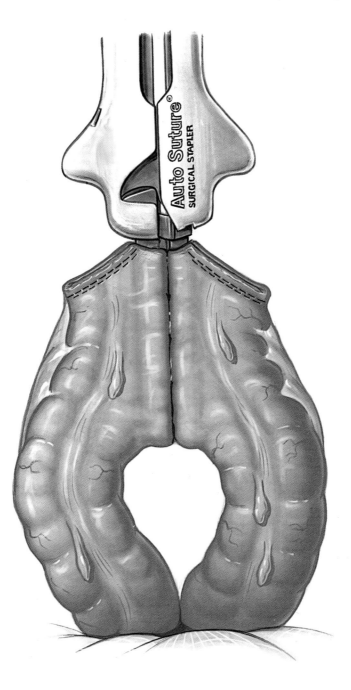

Figure V-2,C: The instrument halves are joined, the bowel ends are evenly aligned, the antimesenteric bowel walls are compressed, and the instrument is locked and activated, creating the side-to-side anastomosis. Following removal of the instrument, the anastomosis is checked to ensure hemostasis.

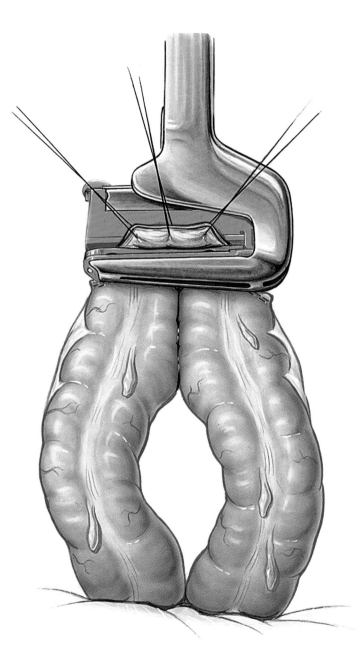

Figure V-2,D: The now common GIA introduction site is closed with a linear instrument, taking care to include the entire circumference of the remaining opening, all tissue layers, and the ends of all staple lines. When closing the instrument, the anastomotic staple lines should be juxtaposed to avoid direct apposition. The excess tissue is excised along the instrument edge.

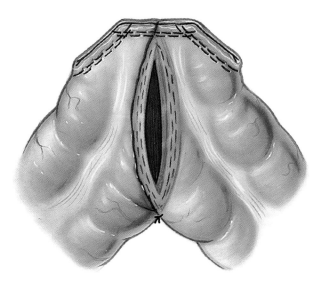

Figure V-2,E: The final result is an oval-shaped anastomotic surface, usually greater than the cross-section of the host bowel lumen. An anchor suture may be placed at the distal end of the anastomosis for additional security.

THE V-SHAPED CLOSURE OF THE GIA INTRODUCTION SITE IN THE FUNCTIONAL END-TO-END ANASTOMOSIS

The geometry of construction of the functional end-to-end anastomosis allows for a variety of solutions that provide flexibility and adaptability to different, sometimes unexpected, intraoperative requirements. An example of this is the variable V-closure of the GIA introduction site, resulting in a "custom made" anastomotic surface (Welter, 1980).

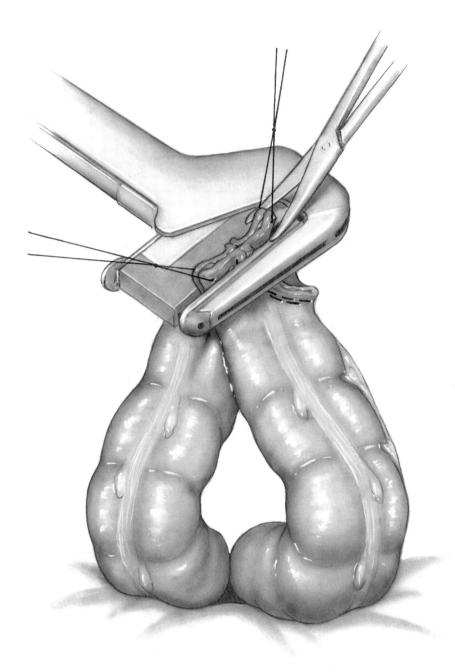

Figure V-3,A: If a more generous anastomotic surface is desirable, the GIA staple lines at the open introduction site can be positioned at a right angle to the plane of the bowel closure lines. The resulting opening is closed with a linear stapler in a cruciate fashion, taking great care to include all tissue layers and to overlap the ends of the previously placed staple lines. The excess tissue is excised.

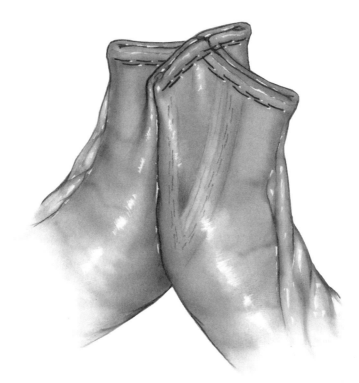

Figure V-3,B: An intermediate V-shaped anastomosis is obtained, enlarging the cross-section approximately one-and-a-half times as compared to the oval anastomotic shape. The crossing of the staple lines, done with care, will heal safely.

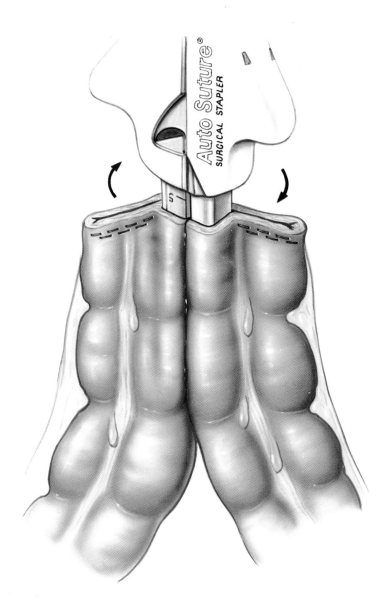

Figure V-3,C: A narrow V-shaped anastomosis can be planned by rotating the forks of the GIA instrument some 45 degrees around their long axis, before closing the instrument halves. In this way, the antimesenteric walls of the bowel ends are slightly shifted vis-à-vis each other at their line of contact.

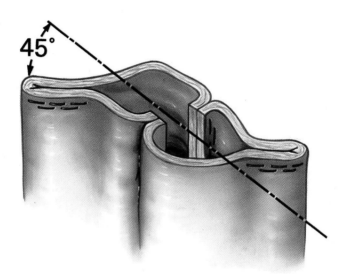

Figure V-3,D: The geometry of this shift is shown.

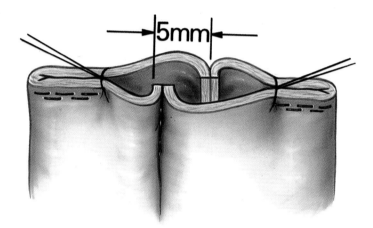

Figure V-3,E: The narrow V-shaped configuration is demonstrated when the open lumen is aligned with the plane of the previous bowel closures.

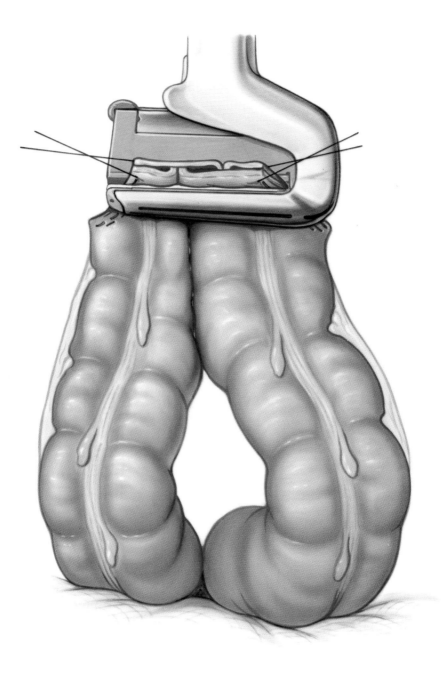

Figure V-3,F: The single, open introduction site is closed in line with the existing bowel closures and the excess tissue is excised as previously shown.

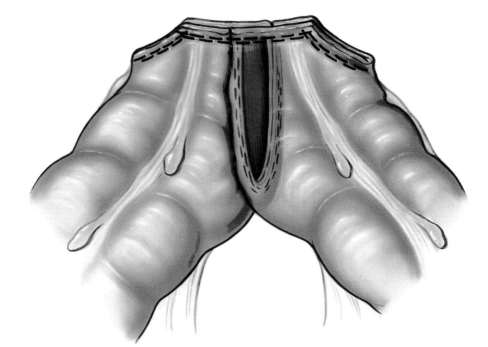

Figure V-3,G: The result is a narrow V-shaped anastomotic cross-section.

"ANASTOMOSE-RESECTION INTEGREE" ANASTOMOSIS FIRST - RESECTION SECOND WITH THE FUNCTIONAL END-TO-END ANASTOMOSIS

The technique of "anastomose-resection integree" (Welter, 1985) can be useful at all levels of the GI and colorectal tracts, especially if a gentle hold on the specimen is desirable to maintain the anastomotic site in a safe and accessible position. This is more obvious in narrow anatomical sites, such as for the cervical esophagogastrostomy or esophagocolostomy, the high gastrojejunostomy, and the various small and large bowel anastomoses in the laparoscopically assisted mode.

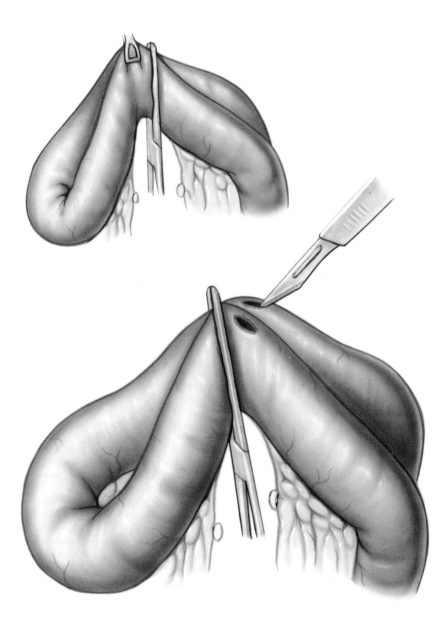

Figure V-4,A: After mobilization, the specimen to be resected is looped, by apposing the proximal and distal points of transection, and held in place with a Kocher clamp placed at a 60-degree angle across both bowel lumina at the transition point between viable and nonviable bowel.

Figure V-4,B: A stab wound is placed into the antimesenteric border of both bowel lumina.

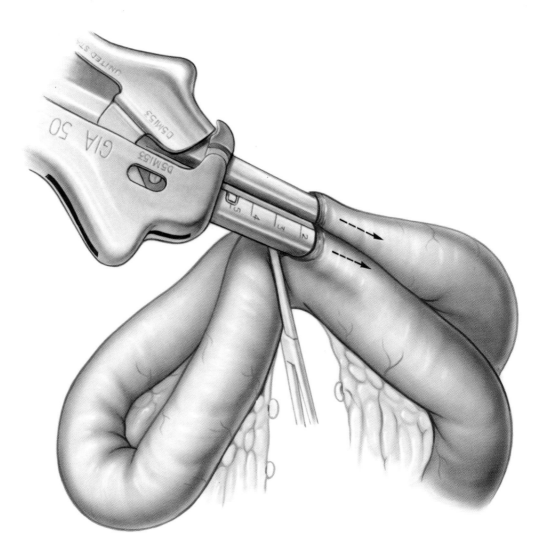

Figure V-4,C: One fork of the GIA instrument is inserted into each lumen, and the instrument is closed and activated, creating the side-to-side anastomosis.

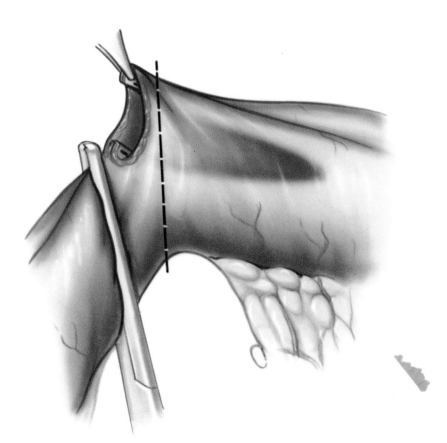

Figure V-4,D: A V-shaped anastomosis is obtained by placing traction on the anterior anastomotic staple line and holding the anastomotic lines in opposition.

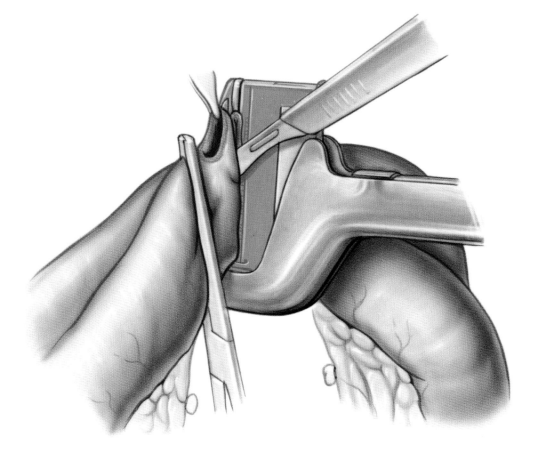

Figure V-4,E: The anastomosis is completed by closure with a linear stapler and resection of the specimen.

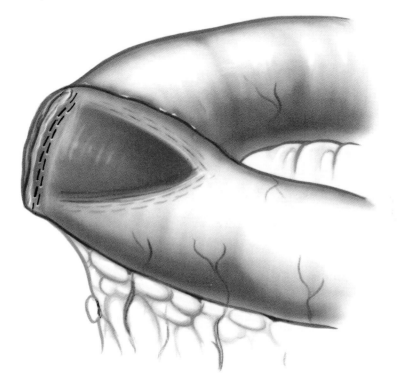

Figure V-4,F: The result is a relatively wide V-shaped anastomotic cross-section.

TRUE SIDE-TO-SIDE ANASTOMOSIS
FOR INTESTINAL BYPASS PROCEDURES

In its best-known application, this procedure was used as an enteroenterostomy, distal to a gastrojejunostomy after subtotal gastrectomy, to avoid an afferent loop syndrome and was known as "Braun's Anastomose" in Germany. In rare instances, it can still be used for that purpose or for any bypass in small and large bowel, where an expeditious execution of the operation is important.

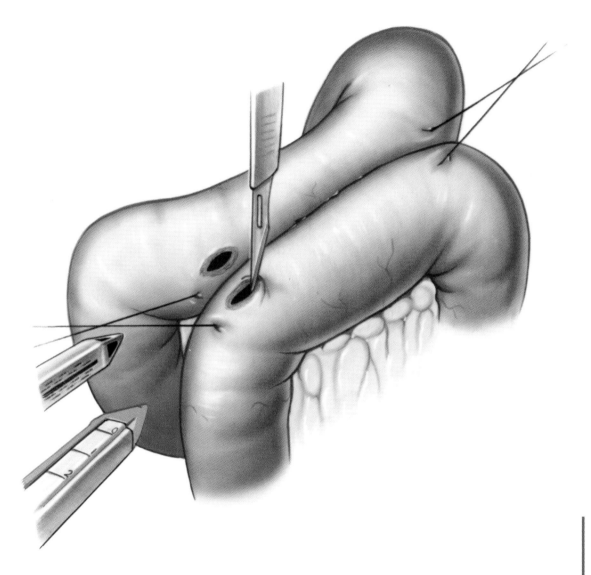

Figure V-5,A: The bowel segments to be anastomosed are approximated with stay sutures, and parallel stab wounds are made in the antimesenteric border of each bowel lumen.

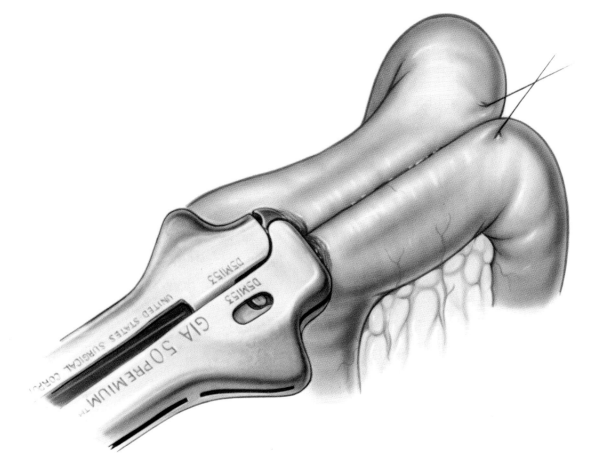

Figure V-5,B: The forks of the GIA instrument are placed into each respective lumen and the instrument is closed and activated, performing the side-to-side anastomosis.

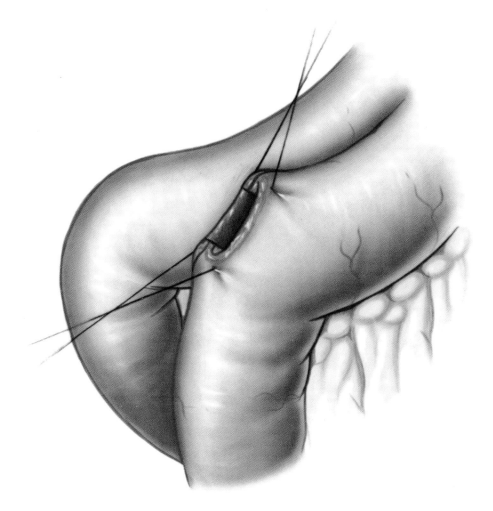

Figure V-5,C: After inspection for hemostasis, the anastomotic staple lines are held in opposition by everting stay sutures or Allis clamps in preparation for closure of the now common opening.

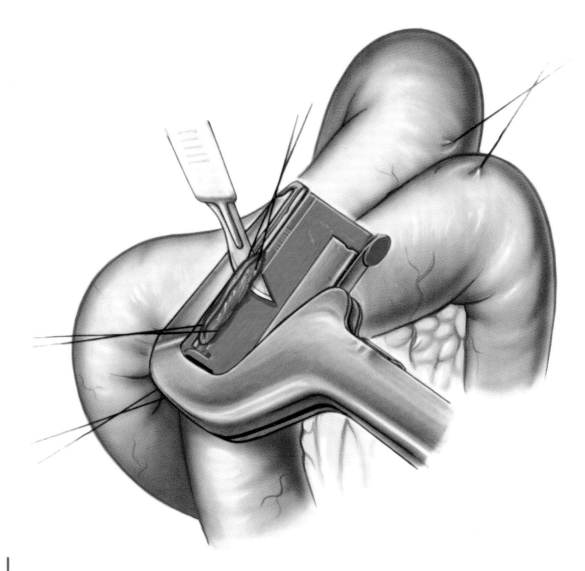

Figure V-5,D: The opening is closed with a TA instrument, ensuring that all tissue layers and the ends of the GIA staple lines are incorporated within the stapler. The excess tissue is excised.

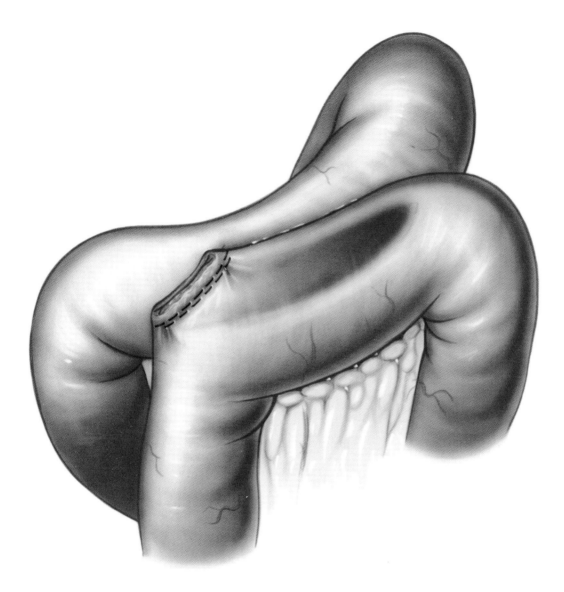

Figure V-5,E: The result is a rapid, large stoma enteroenterostomy.

EXCISION OF MECKEL'S DIVERTICULUM

The technique shown here is best suited to a diverticulum with a narrow opening into the ileum, so as to allow a closure at a right angle to the longitudinal axis of the bowel without encroaching on its lumen and without leaving ectopic gastric mucosa within or beyond the closure. If these conditions cannot be met, then the surgeon is better advised to perform a limited segmental small bowel resection, as shown in Figure V-2, A to E. The use of mechanical sutures renders an incidental diverticulectomy particularly easy and safe.

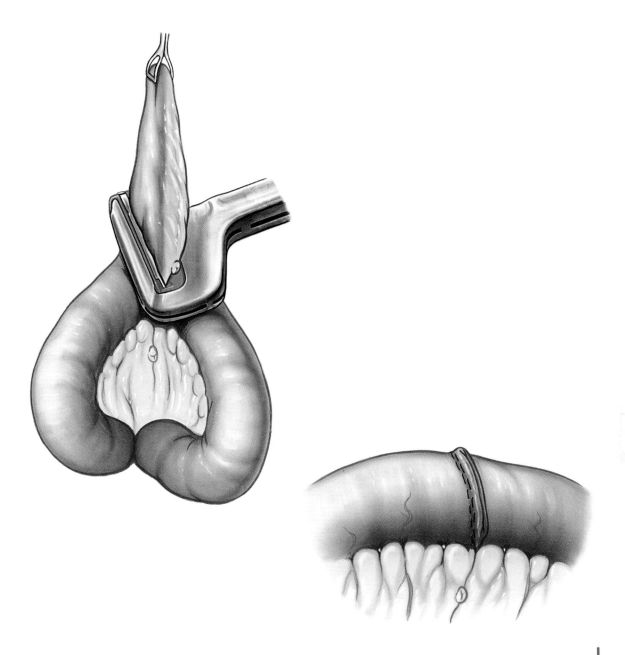

Figure V-6,A: After ligation and division of the vessels to and from the diverticulum, its base is closed with the TA stapling instrument, placed onto normal appearing ileum, at a right angle to the longitudinal bowel axis.

Figure V-6,B: The transverse closure (Mikulicz principle) prevents narrowing of the bowel lumen, by placing the resulting scar at a right angle to the long axis of the intestinal tube.

APPENDECTOMY

The techniques shown are useful in appendectomy for acute appendicitis, as well as for incidental appendectomy; albeit that the technique using the GIA instrument (Figure V-7,B) is particularly suited to the removal of an acutely inflamed appendix, because the cecal and appendiceal lumina are closed simultaneously, avoiding contamination. Both techniques allow the placement of the staple line onto healthy cecal wall, if the inflammation of the appendix extends beyond its base.

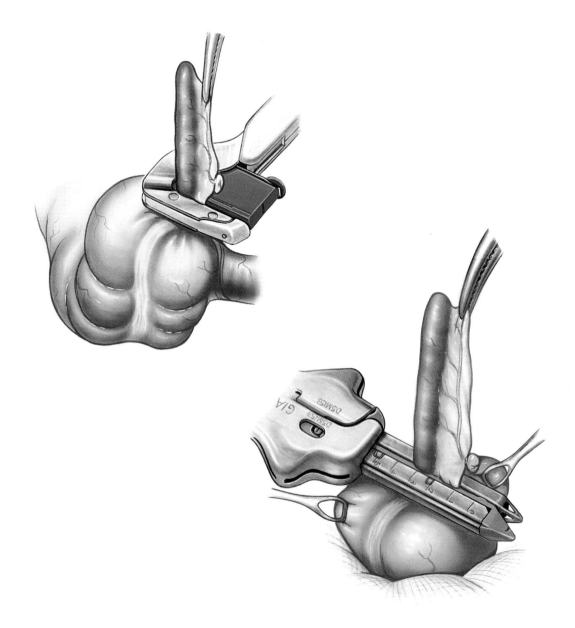

Figure V-7,A: Following ligation and division of the mesoappendix and the appendiceal vessels, the cecum is closed at the base of the appendix with the TA instrument. The appendix is then transected between this instrument and a clamp (not shown here), placed peripheral to the stapler.

Figure V-7,B: Alternatively, the appendectomy can be performed using the GIA instrument; placing staple lines on both the cecum and the appendix, while the instrument's blade transects the cecoappendiceal junction simultaneously. In both instances, the cecal closure may be inverted with a purse-string suture.

RESECTION OF TERMINAL ILEUM, APPENDECTOMY, AND ILEOCECOSTOMY

This technique is applicable to mostly benign conditions limited to the terminal ileum. It requires careful conservation of the blood supply, to and from the cecum, so as to avoid any late anastomotic failures due to vascular impairments that would not be apparent during the procedure. The concern for this possible delayed complication has led to the conventional wisdom that for lesions close to the ileocecal valve, an ileocecal-ascending colon resection with ileo-proximal transverse colon anastomosis is the treatment of choice, especially for malignant conditions where only a complete resection of corresponding mesentery and mesocolon can ensure a satisfactory lymphadenectomy. However, in patients where sufficient ileum for end-to-end ileoileostomy would be available, the technique described here is preferable, since the ileocecal valve represents a functional obstruction distal to the anastomosis that can jeopardize healing.

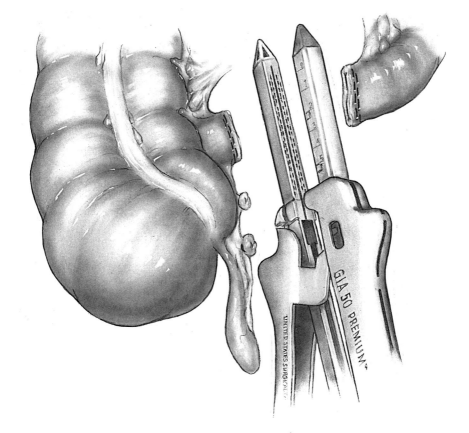

Figure V-8,A: The terminal ileum is resected between two applications of the GIA instrument.

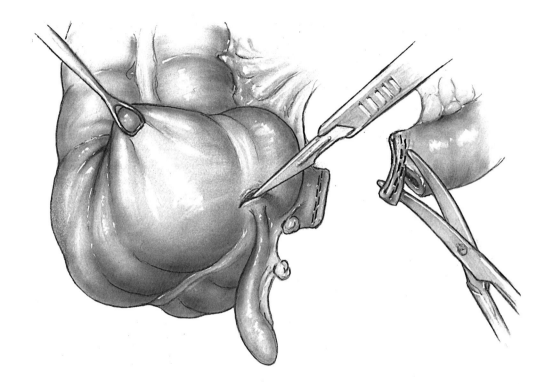

Figure V-8,B: The antimesenteric corner of the proximal ileal staple line is excised, and a stab wound is made into the lumen of the cecum just above the appendicocecal junction.

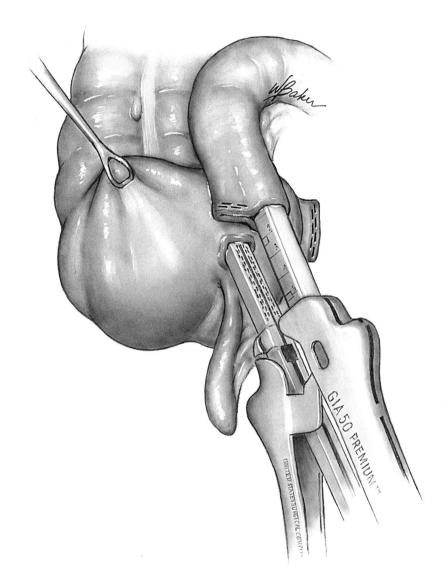

Figure V-8,C: One fork of the GIA instrument is placed into the cecum and one fork into the lumen of the ileum along the antimesenteric border. The instrument halves are joined, the bowel evenly aligned, and the instrument closed and activated, creating the side-to-side ileocecostomy. Hemostasis of the anastomosis is confirmed after removal of the instrument, by inspecting the staple lines.

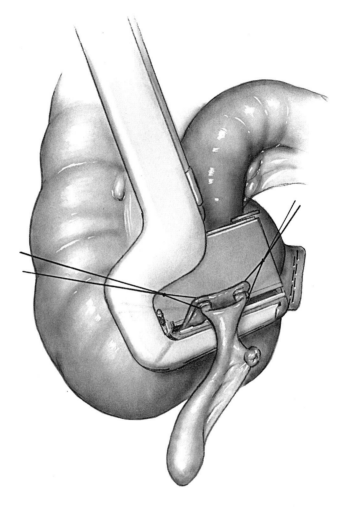

Figure V-8,D: Everting stay sutures or Allis clamps are placed at the ends of the anastomotic staple lines and held up, so as to demonstrate the lateral recesses and borders of the remaining opening. The now common opening is closed below the appendix with the TA instrument, ensuring that all tissue layers are incorporated and the ends of the staple lines are crossed. The excess tissue and appendix are excised.

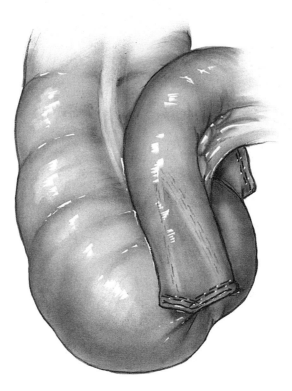

Figure V-8,E: The resulting open V-shaped anastomosis generally exceeds the lumen of the ileum.

COMBINED COLONOSCOPY AND LAPAROTOMY IN THE TREATMENT OF AN ILEOCECAL ANASTOMOTIC FISTULA

In a case of an enterocutaneous fistula that had developed subsequent to a resection of the terminal ileum for Crohn's disease followed by an ileocecal anastomosis, the exact location of the fistula could only be determined by colonoscopy. The combination of endoluminal endoscopy and open operation (as described here) or of endoluminal and endocavitary endoscopy to locate a discrete lesion that can be corrected or remedied by local or segmental excision is one more example of the minimally invasive concept.

The placement of a linear staple line on the soft, healthy bowel wall proximal to the level of the opening from the lumen of the hollow viscus into the fistula accomplishes closure and separation of the bowel lumen from the fistulous tract without encroaching on the diameter of the bowel lumen. By maintaining the linear instrument in position after the staples have been placed, its lateral edge can serve as a guide to the transection of the fistula and the surrounding scar tissue. The fistulous tract with surrounding scarred, inflammatory tissue can then be elevated and dissected towards the periphery and excised, while the healthy bowel is safely maintained deep to this dissection.

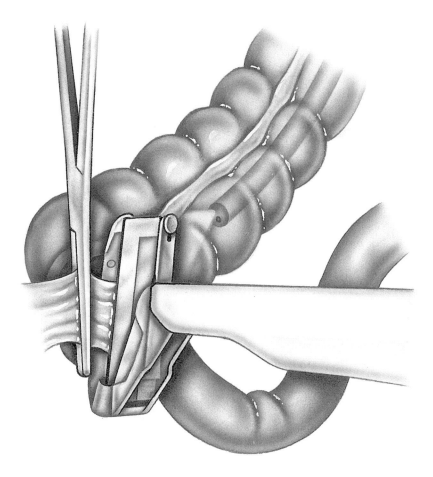

Figure V-9,A: Following colonoscopy to identify the lesion— in this instance, an enterocutaneous fistula— the patient is placed into a modified lithotomy position and given general anesthesia. The abdomen is explored through an appropriate incision and the area of disease— in this case, the right lower quadrant— is dissected to demonstrate the lesion and to initiate surgical repair or removal, as the case may be.

After dissecting the origin of the fistula and its tract to the skin, the TA linear stapler is placed onto the normal bowel wall at the origin of the fistula and is activated, placing a double row of staples on the healthy bowel wall. The fistulous tract is then transected between the stapling instrument and a clamp.

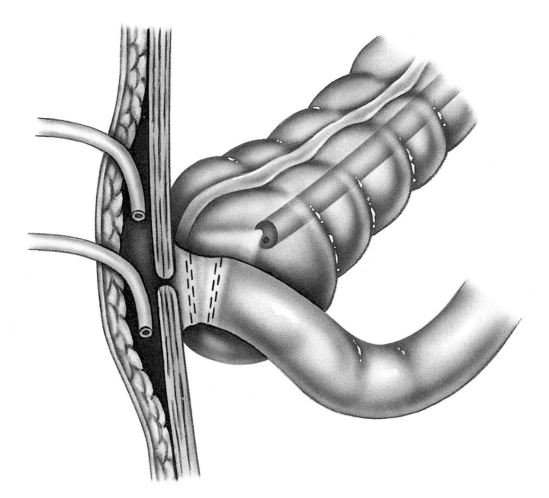

Figure V-9,B: After dissecting the fistula towards the abdominal wall, it is closed with a second application of the linear stapler to avoid later spillage from a subcutaneous abscess cavity into the peritoneal cavity. The fistula, with surrounding inflammatory tissue, is then transected deep to this second linear closure and removed from the operative field. The subcutaneous abscess is drained.

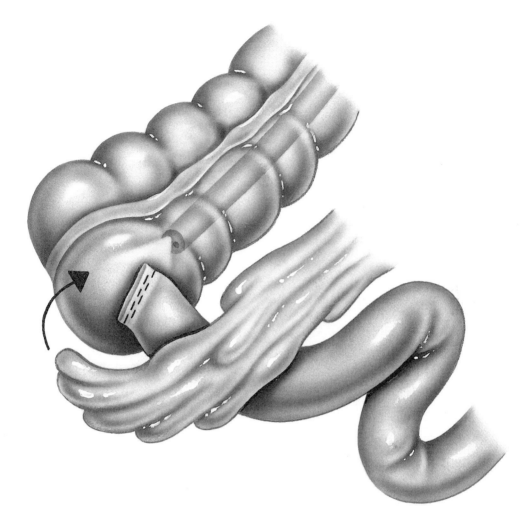

Figure V-9,C: A portion of the omentum is moved towards the bowel closure to serve as a covering pedicle graft.

In patients with a single benign, intraluminal, discrete lesion (e.g., polyp), or a group of them, that does not lend itself to simple excision, the intraoperative use of colonoscopy to localize the lesion(s) is just as helpful as demonstrated in this example of an enterocutaneous fistula. The lesion can then be excised by segmental bowel resection and functional or traditional anastomosis with mechanical suture instruments, either through a laparotomy or a laparoscopic approach.

ILEO-TRANSVERSE COLON BYPASS OPERATION

This procedure is useful in patients who have an incurable, unresectable, obstructing carcinoma of the cecum or ascending colon and whose general condition justifies a palliative procedure, enhancing the quality of remaining survival. Similar bypass operations can be undertaken at various levels of the small and large bowel for comparable indications and purpose.

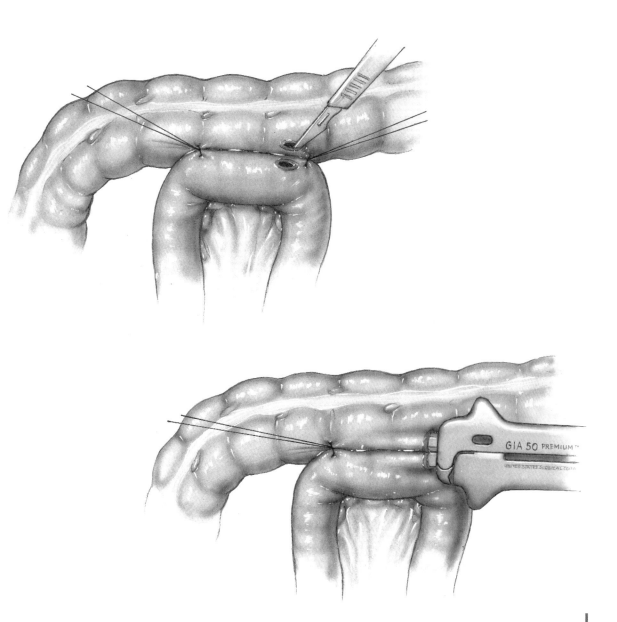

Figure V-10,A: The antimesenteric borders of the terminal ileum and transverse colon that have been chosen for anastomosis are approximated with stay sutures and parallel stab wounds are made into each bowel lumen.

Figure V-10,B: The forks of the GIA instrument are placed into the respective lumina and the instrument is closed and activated, performing the side-to-side anastomosis.

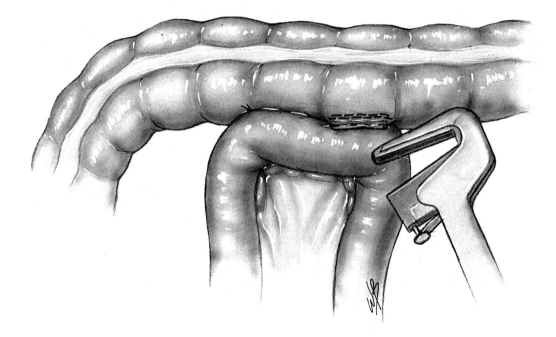

Figure V-10,C: After inspection for hemostasis, the anastomotic staple lines are held in opposition by everting stay sutures or Allis clamps and the common opening is closed with a TA instrument, ensuring that all tissue layers and the ends of the GIA staple lines are incorporated within the stapler. The excess tissue is excised along the instrument edge.

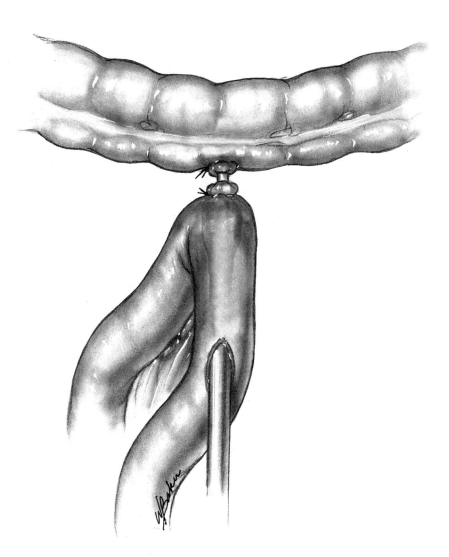

Figure V-10,D: Alternatively, the ileocolostomy can be performed utilizing the circular anastomosing instrument. A 2.5 to 3 cm incision is made along the antimesenteric border of the transverse colon at the site of the proposed anastomosis. A manual purse-string suture is placed around the incision using a monofilament suture. The anvil of the EEA instrument is placed into the lumen of the colon and the purse-string suture is tied securely on the anvil shaft. The site of the anastomosis of the ileum is identified and a 3 cm incision is made in the antimesenteric border at an adequate distance from the anastomotic site. The EEA instrument is introduced through the incision and advanced to the point of anastomosis.

The instrument is positioned so that when opened, the trocar tip will perforate the wall of the ileum on the antimesenteric border. A manual purse-string suture is placed around the perforation to prevent any serosal tearing or tissue slippage as the instrument is closed. This suture is tied snugly but not too tightly around the center rod of the instrument to prevent invaginating tissue within the cartridge as the instrument is closed. The instrument is closed, compressing the ileal and colon walls between the cartridge and anvil, and activated. Following removal of the instrument, the tissue "donuts" remaining within the instrument are examined to ensure that all tissue layers are present in a circular configuration.

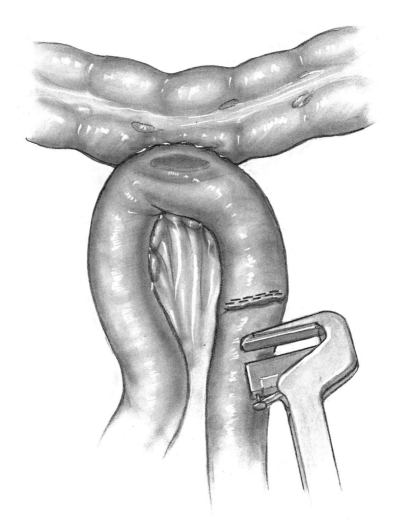

Figure V-10,E: The anastomosis is inspected for hemostasis and the ileotomy is closed transversely with the TA stapler.

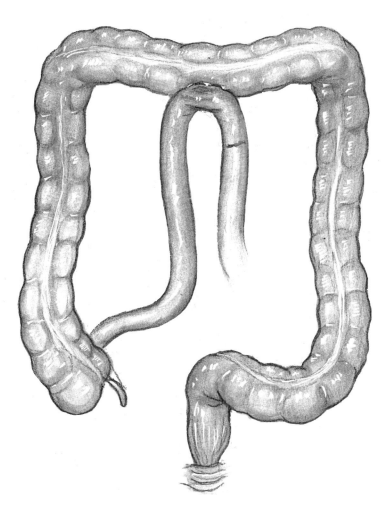

Figure V-10,F: The resulting anastomosis is performed without tension and provides a patent bypass in palliative procedures.

RIGHT HEMICOLECTOMY AND ILEO-TRANSVERSE COLON ANASTOMOSIS

This represents the procedure of choice in patients with benign or malignant lesions in whom an extensive bowel resection, with corresponding, radical mesocolon excision and appropriate lymphadenectomy are indicated. While its intent is curative, it can be used as a palliative procedure in patients whose tumor is resectable and in whom a reasonable life expectancy mandates an optimal quality of life, especially if ever-improving adjuvant treatments are available to reduce or eliminate distant metastases. Even if possible and reasonable from an etiological point of view, a lesser procedure, such as ileocecectomy for lesions near the ileocecal valve or confined to the cecum, is very rarely indicated because of the tenuous blood supply to the ascending colon and even more so to an ileo-ascending colon anastomosis.

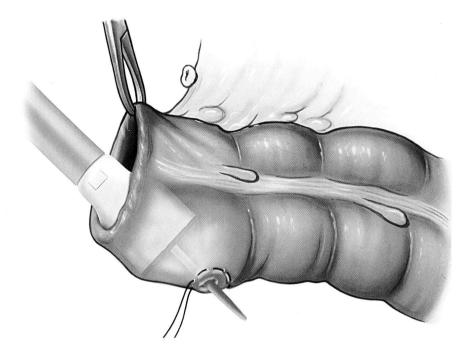

Figure V-11,A: The right hemicolectomy and lymphadenectomy are performed according to current state-of-the-art techniques. The ileum is transected between the application of a Purstring instrument and suture on the proximal ileum and a bowel clamp distally. The ileum is sized to ascertain the appropriate EEA instrument to be used to perform the anastomosis. In all cases where the EEA instrument is used, the size chosen should accommodate the bowel lumen snugly to avoid any bunching or pleating of tissue at the site of the anastomosis; however, to avoid any possibility of serosal tearing and subsequent leakage, it should not stretch or exceed the natural lumen diameter. The EEA stapler, without anvil, is inserted into the open transverse colon and positioned at the antimesenteric border, 4 to 5 cm distal to the open bowel rim. The instrument is opened fully allowing the tip to perforate the wall of the colon. A manual purse-string suture is placed around the perforation site to prevent any serosal tearing or tissue slippage, and the suture is tied snugly around the center rod of the instrument.

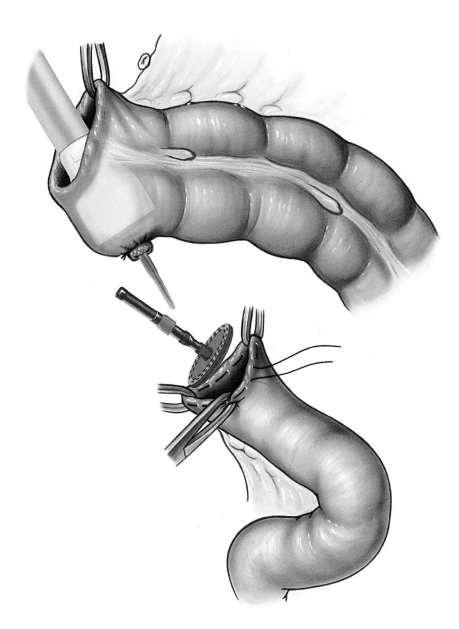

Figure V-11,B: The anvil is inserted into the ileum and the purse-string suture is tied securely on the anvil shaft. Ensure that the tissue is snug against the cartridge and anvil to reduce the possibility of any bunching or overlapping of tissue as the instrument is closed.

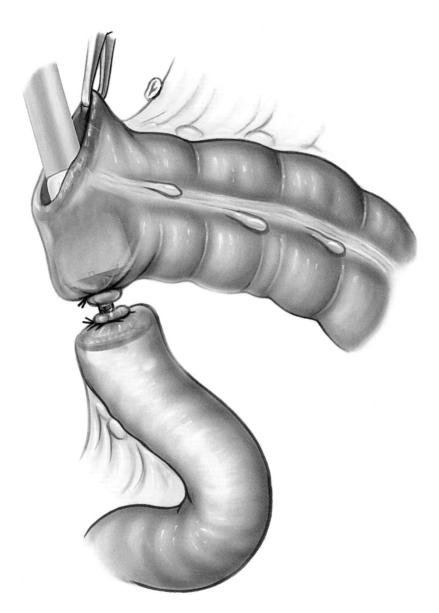

Figure V-11,C: The anvil shaft and center rod of the instrument are joined. The instrument is closed, checked to make sure that no extraneous tissue is incorporated, and activated. The instrument places a circular, double staggered row of staples and cuts the ileocolic stoma. The excised tissue "donuts" remaining within the instrument are examined to ensure that all tissue layers and purse-string sutures are present in both circular remnants.

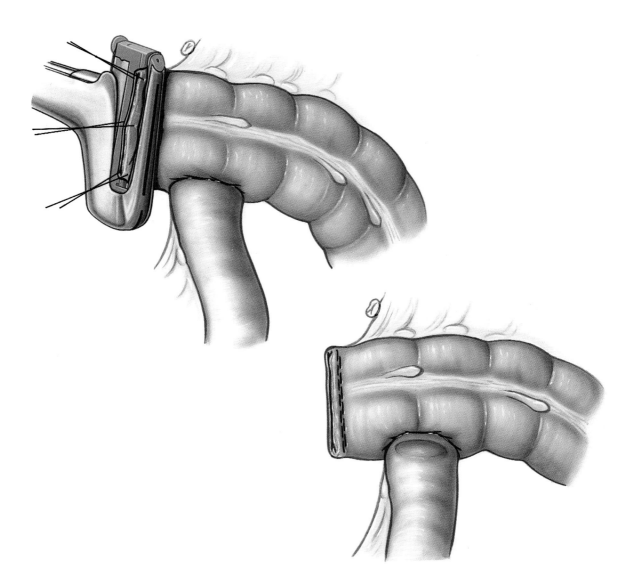

Figure V-11,D,E: The anastomosis is inspected for hemostasis and the transverse colon is closed with the TA stapler, allowing sufficient distance between the anastomosis and this closure for adequate blood supply to the tissue remaining between both.

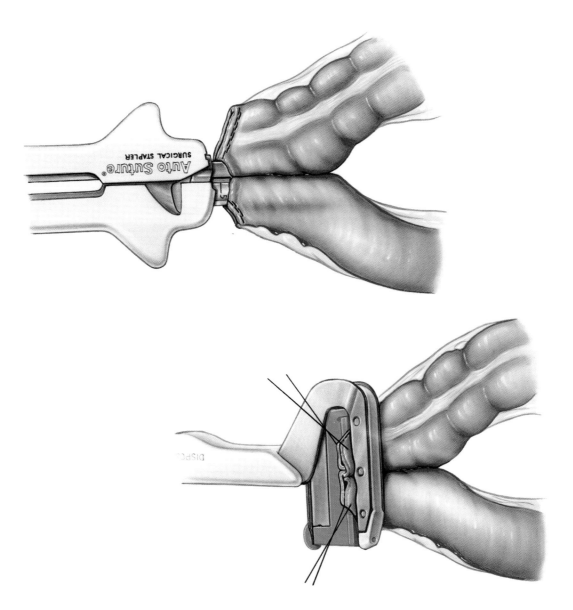

Figure V-11,F,G: The ileocolostomy can be performed utilizing the GIA and TA instruments in a functional end-to-end mode. The functional end-to-end procedure lends itself favorably to joining bowel ends of unequal caliber.

END-TO-END LARGE BOWEL ANASTOMOSIS

This technique represents a close reproduction of the manual, one or two layer(s) anastomosis. Designed in the late 1960s by Turi Josefsen, studied and perfected by her and Gershon Efron in the laboratory, and then used clinically, it filled a void long before the advent of the circular EEA instrument for surgeons who were reluctant to accept the functional end-to-end anastomosis after its introduction in 1968. It continues to be favored by surgeons who prefer a more progressive transfer from traditional manual methods to mechanical techniques and who are fully aware that its construction requires careful attention to detail and avoidance of inherent pitfalls. When all the details and safety measures are respected, the final result is a secure and reliable, widely patent anastomosis.

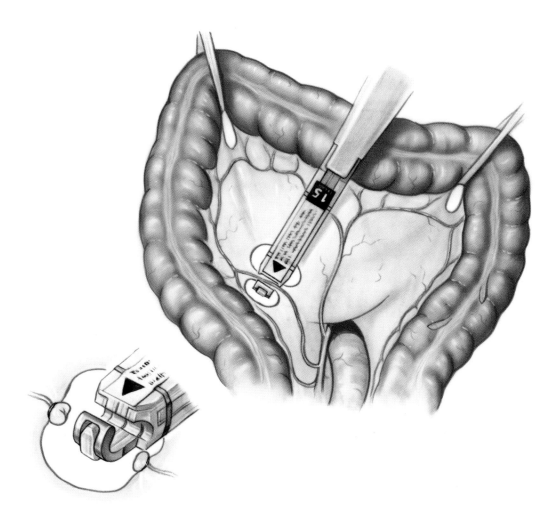

Figure V-12,A: Mobilization of the colon to be resected can be facilitated with use of the PLDS instrument. The mesentery to be ligated and divided is placed into the jaw of the instrument. When activated, the instrument forms two crescent-shaped staples and evenly cuts the tissue held between them. The instrument is used sequentially to complete the mobilization. The colon resection is performed in a manner that satisfies the therapeutic demands and concerns inherent to each individual diagnosis and operative findings.

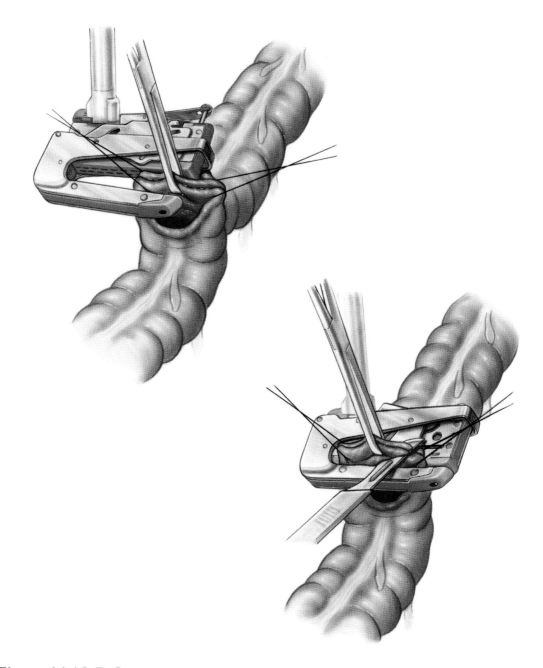

Figure V-12,B,C: The anastomosis is created by following the principle of triangulation with the posterior walls inverted and the anterior and lateral walls everted. The posterior wall edges of the open bowel segments are approximated serosa-to-serosa (inverting) with stay sutures, placed at both ends of approximately half the bowel circumference. An Allis clamp may be placed at the midpoint of these matched posterior walls to ensure incorporation of all tissue layers into the instrument. The linear stapler is placed on the merged posterior bowel edges, beneath the stay sutures and Allis clamp; it is closed and activated. The excess tissue is excised along the instrument edge, leaving the stay sutures intact. This staple line forms the base of the triangle.

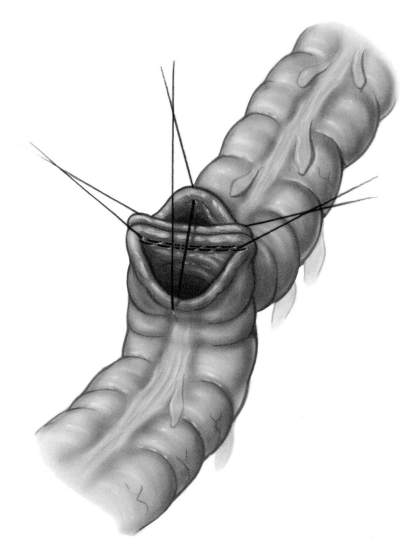

Figure V-12,D: The circumference of the anterior-lateral walls is divided into two equivalent halves by a stay suture, approximating the bowel borders or edges at this level in an everting fashion.

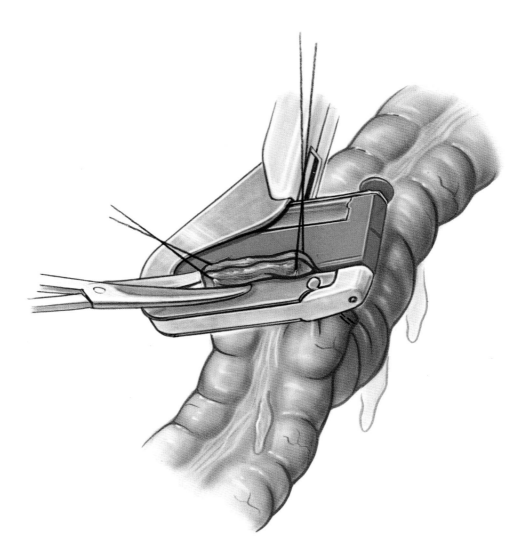

Figure V-12,E: Each half of the anterolateral walls or each anterolateral side of the triangle is closed with the TA stapler, placed beneath a corresponding pair of stay sutures. On this second and third application of the TA instrument, care is taken to include all tissue layers and the ends of the previously placed staple lines. This ensures a secure completion of the circle. The excess tissue is excised.

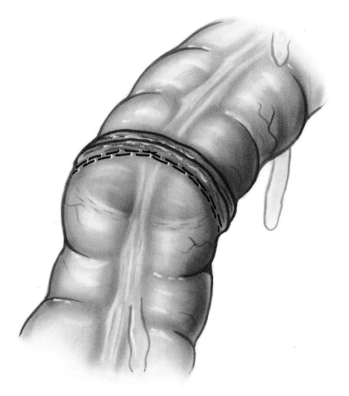

Figure V-12,F: With only the posterior wall of the anastomosis inverted, the patency of the resulting end-to-end anastomosis is excellent.

END-TO-END SMALL BOWEL ANASTOMOSIS

The mobility of the small bowel and ease of rotation around its long axis facilitate the construction of this triangulating, mucosa-to-mucosa (everting) technique, based on the historical experience of Travers (1812) and the contemporary experimental and clinical evidence that mucosa-to-mucosa bowel closures and anastomoses heal safely when known precautions (blood supply, water, and air seal) are respected.

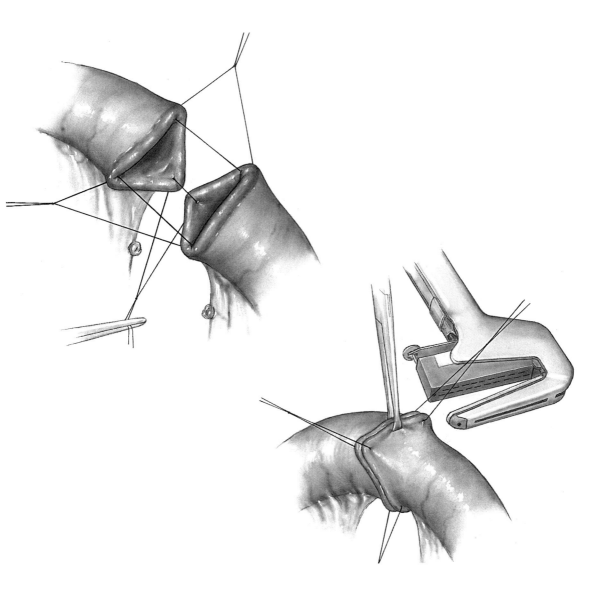

Figure V-13,A: Following resection of the specimen, the edges of the open bowel are approximated with three equidistant stay sutures. The first one is placed to align the mesentery, and the other two sutures share equally the remaining circumference of the bowel.

Figure V-13,B: Two stay sutures are pulled in opposite directions at a time, and an Allis clamp is placed at the midpoint of the pouting bowel edges to ensure incorporation of all tissue layers in the instrument. The TA instrument is placed around and beneath the bowel edges, held in approximation by the sutures and the clamp. To avoid including the opposite wall in the instrument, traction is placed on the third suture as the instrument is closed and activated.

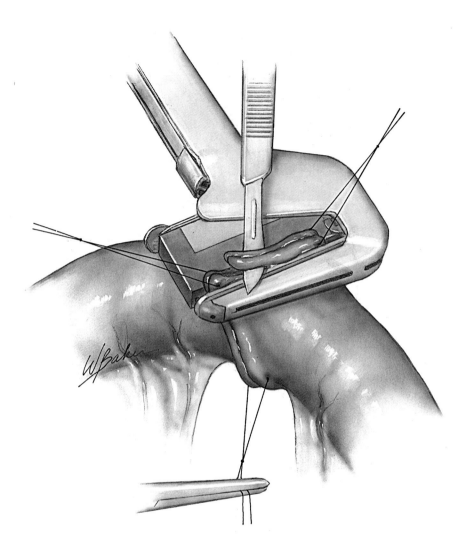

Figure V-13,C: The excess tissue is excised between the stay sutures, which are left in place temporarily to assist with the triangulation.

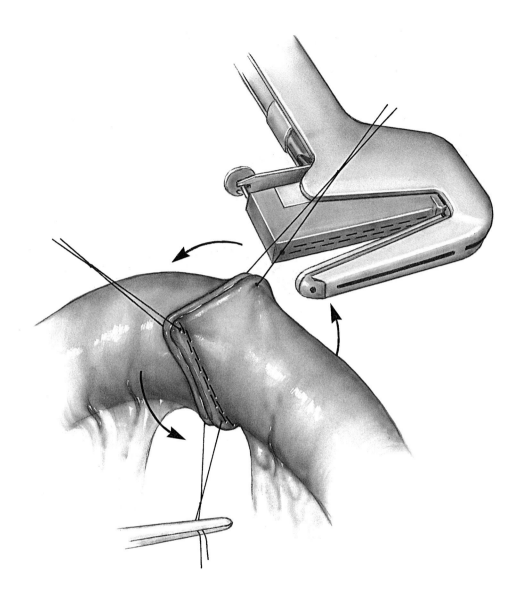

Figure V-13,D: The bowel is rotated and the next side of the triangle is closed taking care to overlap with the end of the previous staple line.

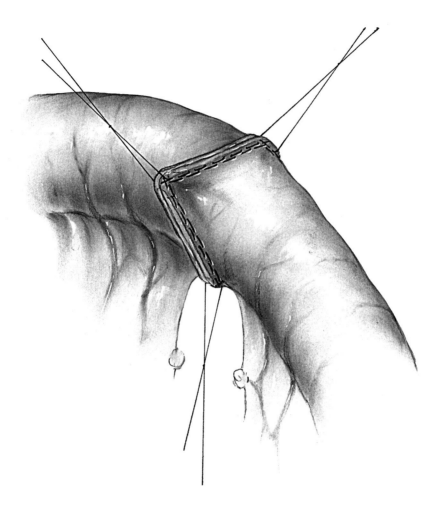

Figure V-13,E: In order to position the final side of the triangle for easy application of the stapler, the first stay suture is passed through the rent in the mesentery, and the third side is closed taking care to include the ends of the previous two staple lines at each extremity of this third staple line.

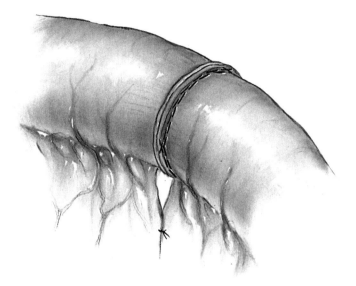

Figure V-13,F: As there is no tissue inversion, the patency of the small bowel is of the same caliber as the normal remaining bowel.

COLOCOLOSTOMY WITH THE CIRCULAR ANASTOMOSING INSTRUMENT

Since the advent of the EEA circular anastomosing instrument in 1977, the end-to-end, inverting anastomosis has become the procedure of choice for the reconstruction of bowel continuity because it reproduces most closely the anatomical features of the traditional, hand-sewn anastomosis. Furthermore, it soon became obvious that its use facilitated joining different hollow visceral structures at various levels of the gastrointestinal and colorectal tracts and in fact reduced anastomotic leaks in esophageal and rectal anastomoses significantly. To reduce the potential of post-operative stricture formation, when preparing the bowel ends prior to anastomosis, care should be taken to clear sufficient mesentery from the bowel walls to prevent incorporation within the anastomosis while still ensuring viability. The distance will vary from 0.25 to 1.5 cm, depending on the diameter size of the EEA instrument chosen.

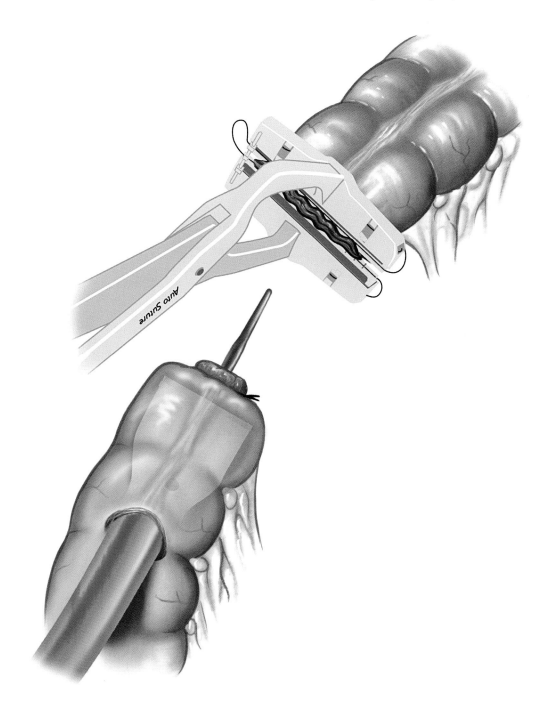

Figure V-14,A: Following mobilization of the specimen, as well as the corresponding mesocolon and lymph nodes, a Purstring instrument is placed on the healthy, remaining bowel sides both proximally and distally, parallel clamps are placed on the specimen side, and the specimen is resected along the edge of the Purstring instruments. The smaller lumen is sized to ascertain the appropriate EEA instrument to be used for the anastomosis. The instrument, without anvil, is introduced into the proximal colon through a colotomy performed approximately 4 to 5 cm proximal to the anastomotic site. The EEA instrument is advanced to the level of the anastomosis, opened fully, and the purse-string suture is tied snugly but not too tightly around the center rod of the instrument allowing the tissue to slide as the instrument is closed.

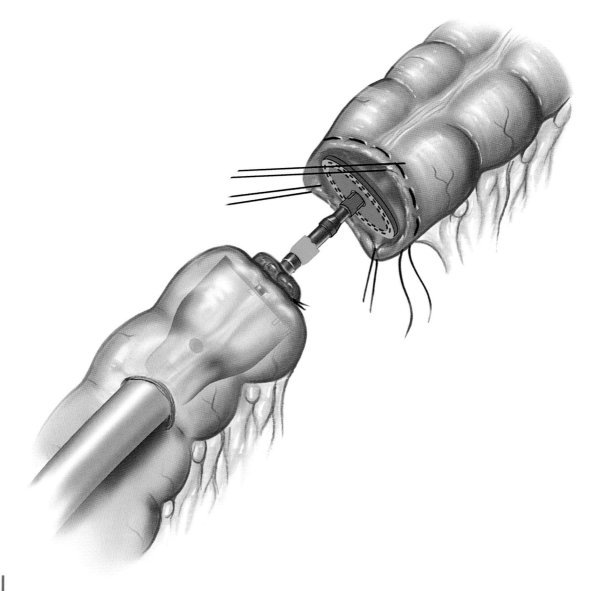

Figure V-14,B: The distal Purstring instrument is removed and three stay sutures or Allis clamps may be placed equidistantly just beyond the purse-string suture to facilitate anvil placement. The anvil is inserted into the colon, and the purse-string suture is tied securely.

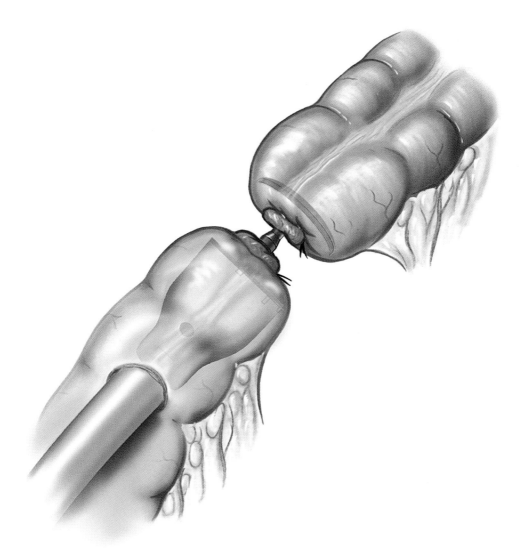

Figure V-14,C: The anvil shaft and center rod of the instrument are mated and the instrument is closed, ensuring that the tissue is snug against the anvil and cartridge and that no extraneous tissue is incorporated.

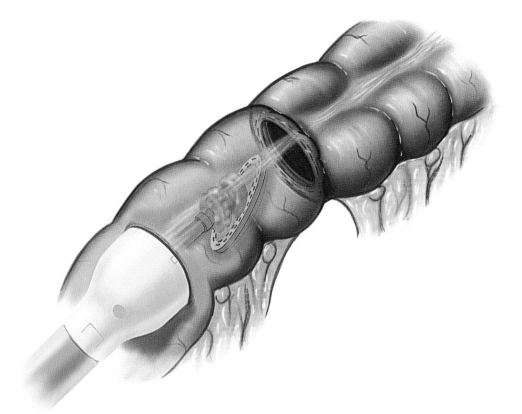

Figure V-14,D: The instrument is activated, opened after the staples have been formed, and then removed. The excised tissue "donuts" remaining within the instrument are checked to ensure that all tissue layers and purse-string sutures are intact. The anastomosis is inspected for hemostasis.

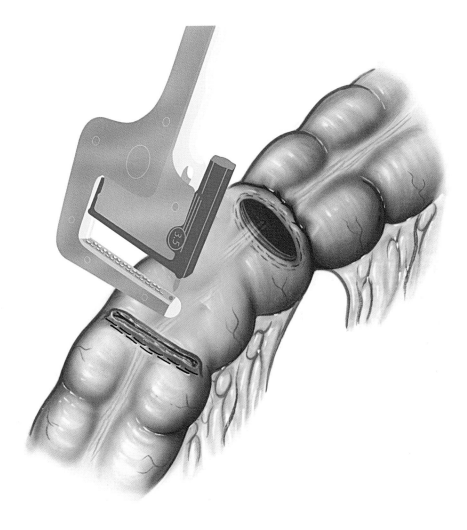

Figure V-14,E: The colotomy is closed transversely with the TA stapler.

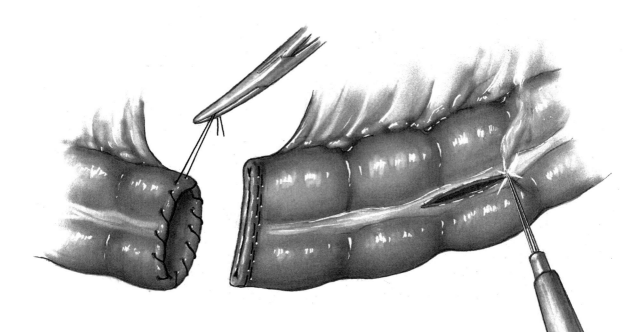

Figure V-14,F: In the event of discrepancy in lumen size between the proximal and distal colon, such as in megacolon, the anastomosis can be performed through a linear staple line. The distal colon is prepared with a purse-string suture, which may be placed manually or with the Purstring instrument, and the proximal colon is closed with the GIA or TA stapler. A colotomy is performed 3 to 4 cm distant from the stapled closure.

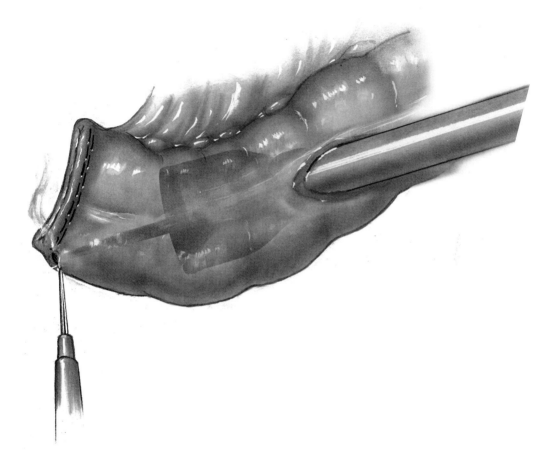

Figure V-14,G: The EEA instrument, without anvil, is introduced into the colon through the colotomy and positioned on the antimesenteric border at the site of the anastomosis. The instrument is opened fully allowing the tip to perforate the wall of the colon. A small opening may be made to facilitate this step or the antimesenteric corner of the staple line may be excised.

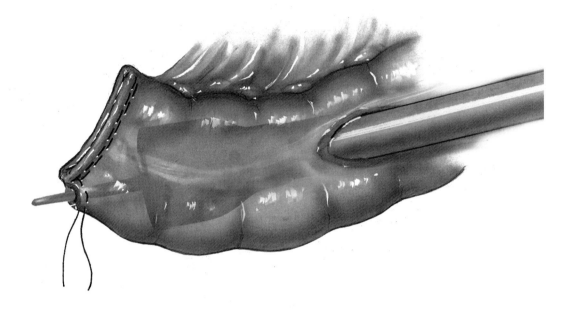

Figure V-14,H: A manual purse-string suture is placed around the perforation site to prevent any serosal tearing or tissue slippage, and the suture is tied snugly but not too tightly around the center rod of the instrument.

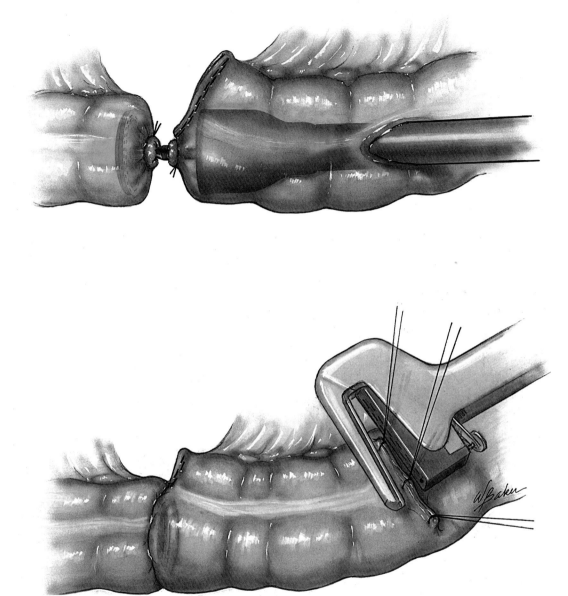

Figure V-14,I: The anvil is inserted into the lumen of the distal colon, and the purse-string suture is tied. The anvil shaft and center rod are joined and the instrument is closed and activated, creating the colocolostomy.

Figure V-14,J: The colotomy wound edges are approximated with sutures or Allis clamps, and the colotomy is closed with the TA stapler. The excess tissue is excised.

COLOSTOMY

With this technique, the open bowel circumference is anastomosed to the flat surface of the skin, resulting in the immediate "maturing" of the colostomy or ileostomy, as the case may be. In patients who have benefited from a planned operation and a regular bowel preparation preoperatively, the colostomy can be placed before the abdomen is closed. In patients who have undergone an emergency operation, with questionable, accelerated bowel preparation, the construction of the colostomy can progress to the level shown in Figure V-15,E. While an assistant is holding the EEA instrument, the surgeon closes the incision and seals it from the operative field. The instrument is only then activated, opening the bowel, without contaminating the closed and sealed incision.

Figure V-15,A: A purse-string suture is placed on the open colon circumference. This may be done with a Purstring instrument, an over-and-over "whipstitch", or an in-and-out suture, parallel to the free edge of the bowel, as shown here. Purse-string sutures should be placed no more than 2.5 mm from the cut edge to avoid excessive tissue within the closed anvil and cartridge. The diameter of the bowel is measured to choose the appropriate EEA instrument. The anvil is separated from the instrument and inserted into the lumen of the colon, and the purse-string suture is tied securely in the notch on the anvil shaft.

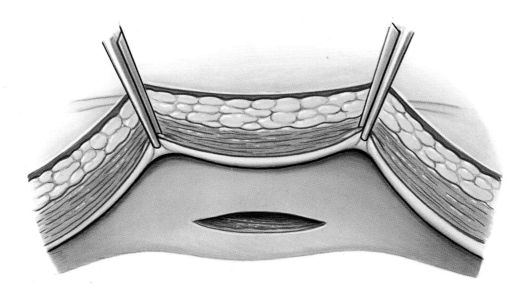

Figure V-15,B: The colostomy site is selected and the peritoneum, muscle, fascia, and subcutaneous tissue to the level of the dermis are incised sufficiently to accommodate the colon.

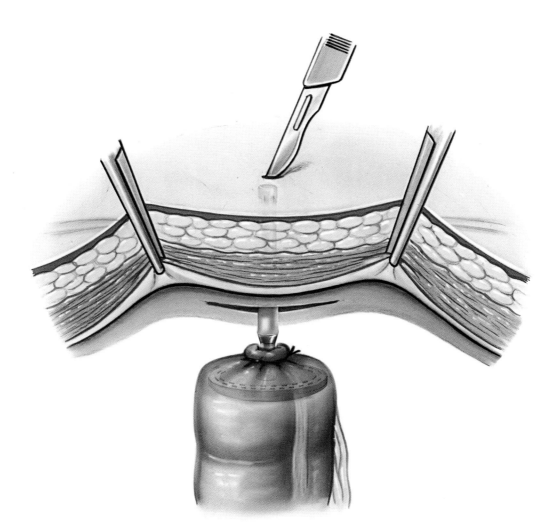

Figure V-15,C: The anvil shaft is introduced through this incision to just beneath the skin, and the skin is incised over the tip of the shaft.

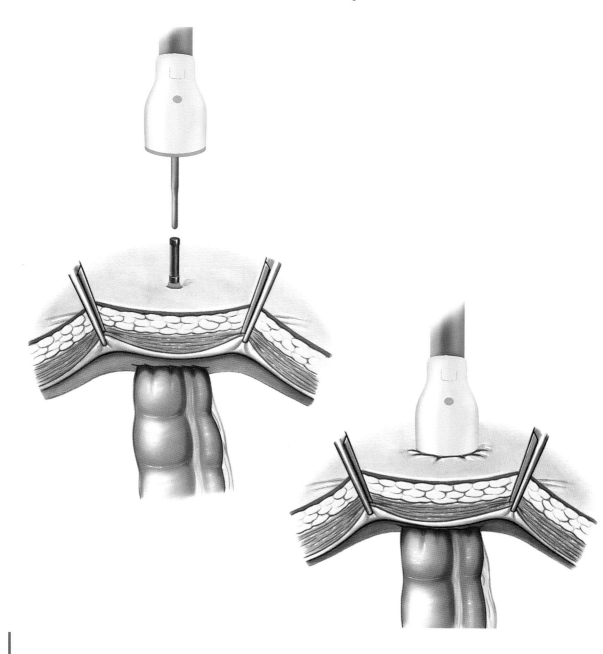

Figure V-15,D,E: The anvil shaft is passed through the skin incision, and the anvil carrying the colon is advanced through the abdominal wall opening, from inside the abdomen to just beneath the skin. The anvil shaft and center rod of the instrument are mated and the instrument is closed and activated to perform the colostomy.

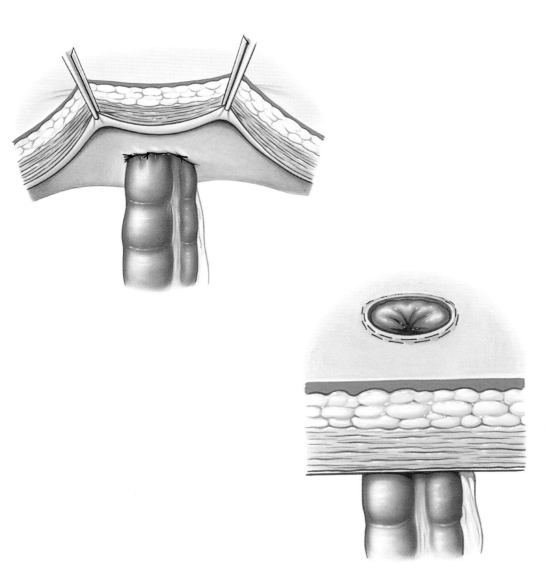

Figure V-15,F: In order to avoid tension on the enterocutaneous anastomotic site, the circumference of the bowel is secured to the peritoneum with a few sutures at the level of its passage into the abdominal wall tunnel.

Figure V-15,G: The EEA instrument helps in the creation of a clean, secure colostomy. The resulting ostium can be influenced by and adapted to a variety of anatomical factors, as well as age and patient habitus, by choosing the appropriately sized instrument out of five available EEA staplers.

ILEOSTOMY

The construction of an ileostomy of normal caliber is usually achieved with the smaller sized circular instruments. The basic technique follows the same steps as previously shown for a colostomy.

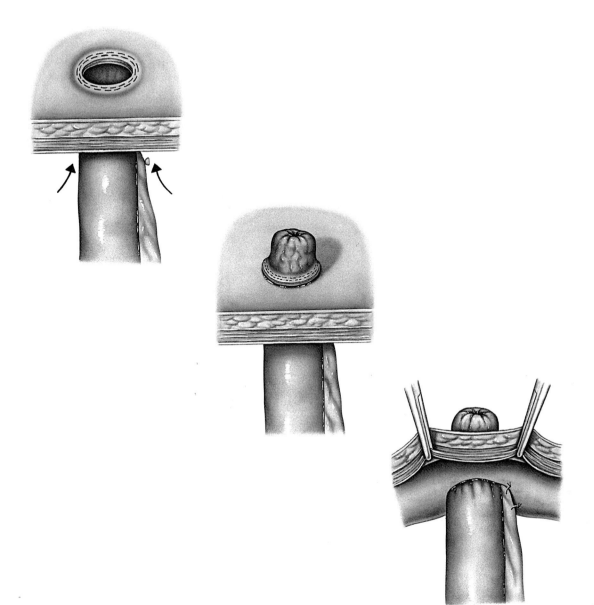

Figure V-16,A: After a secure, circular enterocutaneous anastomosis has been accomplished, the small bowel is pushed from the abdomen through the stoma to create the desired ileostomy nipple.

Figure V-16,B: After the desired amount of bowel has been everted, the nipple is inspected to make sure that it is symmetrical over its entire circumference. Major retractions are corrected by everting adequate amounts of bowel to complete a uniformly even nipple.

Figure V-16,C: The nipple is then anchored in this position with sutures from the bowel wall to the peritoneal opening, placed through the open abdomen.

VENTED SIDE-TO-SIDE COLOCOLOSTOMY

This procedure is inspired by the first two stages of the three-stage Mikulicz operation, modified to the extent that the diseased bowel is excised at the time of the primary operation and not in a second stage and that the stapling instruments make it possible to achieve stages one and two of the Mikulicz procedure during the single, primary operation. The third stage of the original operation, e.g., closure of the common colocolostomy stoma, becomes now an easy second stage. In essence, this is a two-stage functional end-to-end anastomosis, where the GIA stapler introduction site is closed secondarily, after it has been used as a stoma to the skin for drainage of colon contents, to protect the side-to-side colocolic anastomosis and to favor healing of the bowel. Except for the presently rare occasion when a truly diverting colostomy is indicated, this procedure has replaced the venerable Hartmann operation, with its often technically challenging second stage. It is useful in emergency resections, when feasible and safe, of both inflammatory and malignant lesions of the large bowel, where a satisfying preoperative bowel preparation could not be accomplished and where the added step of freeing the splenic flexure in left-sided lesions neither compromises patient survival nor leads to procedure-related intraperitoneal complications. The side-to-side anastomosis facilitates the joining of bowel ends of different caliber, e.g., proximal dilatation due to obstruction, distal normal or collapsed bowel.

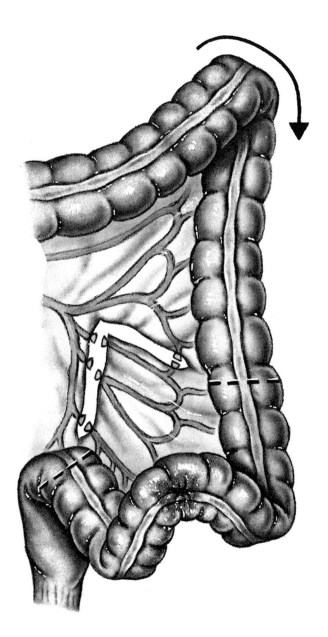

Figure V-17,A: For a left-sided colon lesion, the splenic flexure is dissected free so as to allow sufficient mobility for the side-to-side joining of the remaining bowel ends. This added step should in no way compromise a state-of-the-art resection or excision, as indicated by the particular diagnosis and operative findings, nor should it threaten patient survival or enhance the potential for postoperative intraperitoneal complications. If any of these untoward factors are feared, the patient should receive a time- and battle-proven Hartmann procedure after adequate excision of the offending lesion or even after only a fully diverting proximal colostomy, with plans and preparations to return under more favorable circumstances.

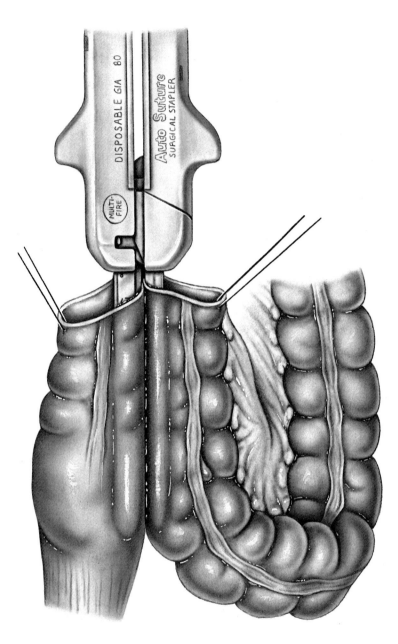

Figure V-17,B: After excision of the involved bowel—in this case, the sigmoid colon—and if a malignancy is present the corresponding mesocolon, the proximal and distal bowel ends are joined side-by-side and anastomosed with the longer GIA instrument. In this illustration, the anastomosis is performed through open bowel ends; however, under emergency circumstances, the sigmoid colon would have been resected between two applications of the GIA stapler so as to avoid contamination. The side-to-side anastomosis would then be performed as shown in Figure V-2, A to C.

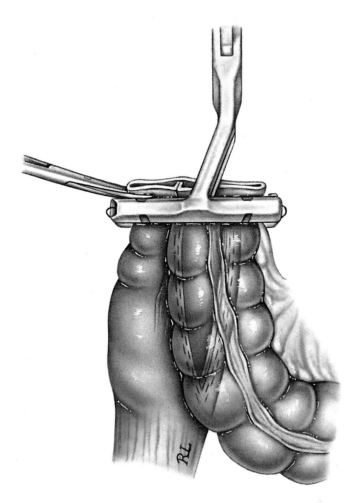

Figure V-17,C: A purse-string suture is placed around the common stomal circumference.

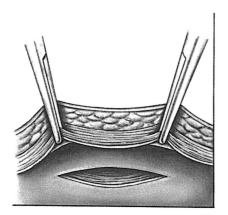

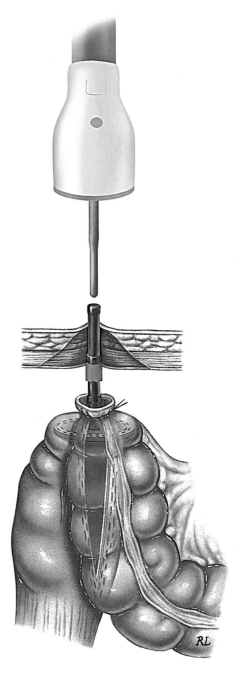

Figure V-17,D: The peritoneum, muscles, fascia, and subcutaneous tissues are incised up to the dermis from inside the abdomen, at the site selected for the placement of the ostomy.

Figure V-17,E: The anvil of the circular anastomosing instrument is placed into the common opening of afferent and efferent colon. The purse-string suture is tied around the anvil shaft, and the end of the shaft is pushed against the skin from inside the abdomen and out through a small incision made above the tip. The anvil shaft is attached to the center rod of the EEA instrument.

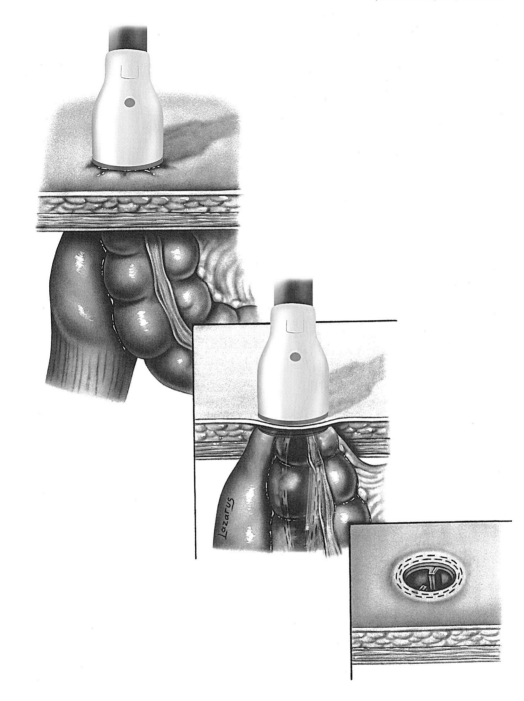

Figure V-17,F,G,H: The instrument is closed and activated, and the colocutaneous anastomosis is accomplished. While the discharge from the stoma is initially copious, it diminishes progressively over four to five days postoperatively. As the side-to-side anastomosis re-establishes a normal route, the secretion from the colostomy is reduced to minor amounts of mucus at the end of one week to ten days.

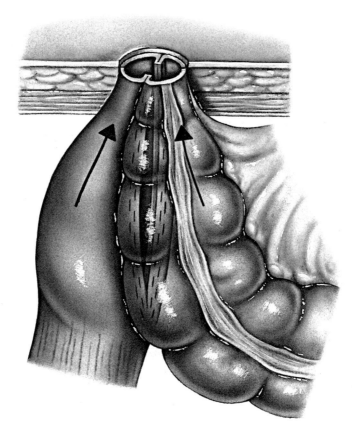

Figure V-17,I: The stoma heals and contracts to a small opening. The second stage of the operation— the closure of the stoma— can usually be undertaken within a month after the procedure. It is invariably possible to proceed under local anesthesia, after satisfactory bowel preparation. The facilities available in the average ambulatory care center are usually satisfactory.

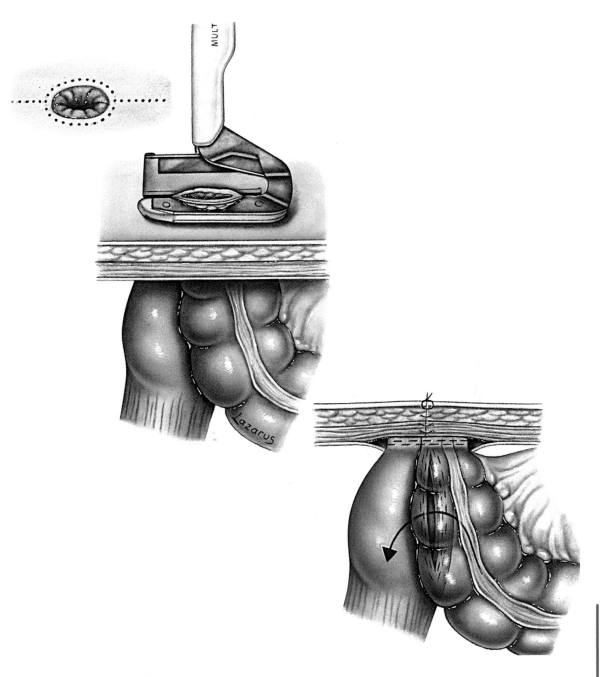

Figure V-17,J,K,L: After local anesthesia has been obtained, the abdominal wall is incised around the bowel edges, down to the peritoneum. The abdomen is not entered. The stoma is closed with a linear stapler, the muscles are re-approximated over this closure, and the subcutaneous wound is packed and allowed to heal by secondary intention or delayed closure, as circumstances indicate.

CURE OF CECAL VOLVULUS

Cecal volvulus due to "cecum mobile," the result of incomplete bowel rotation during fetal life, is always acute in its obstructive manifestations, as well as threatening to organ integrity, if the volvulus is not reduced and vascular patency is not reestablished within reasonable limits for tissue survival. Often though it is encountered as a chronic presentation in patients who have learned to cope with their recurring predicament, either through self-taught maneuvers or by seeking medical help that is effective short of a major procedure. Whatever the presentation, when and if such patients seek a definitive cure and their cecum is viable, the surgeon can usually reduce it with ease but is then faced with an enlarged, thickened bowel segment that does not fit into its usual anatomical bed. If the cecum is gangrenous, an ileo-cecal-ascending colon resection has to be performed. But rather than subjecting the patient with a viable cecum to the sacrifice of the ileocecal valve in order to accommodate the remaining bowel to the available space, a tailoring operation with the GIA instrument is feasible in both acute and chronic recurrent volvuli.

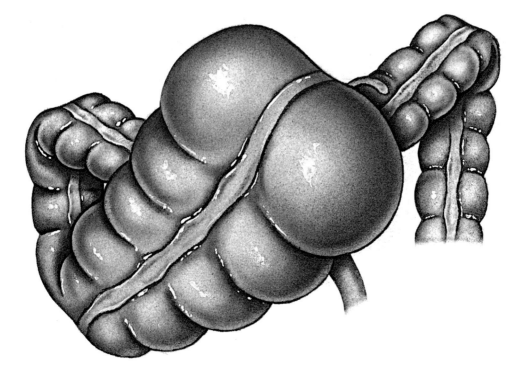

Figure V-18,A: The axis of the twist or volvulus is usually transverse, at a right angle to the long axis of the cecum, just below the hepatic flexure. Since the terminal ileum is also unattached to the posterior peritoneum, the cecum and ileum flip superiorly and medially.

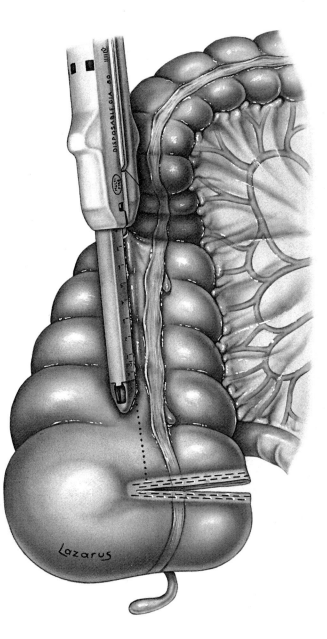

Figure V-18,B: Reduction of the viable cecum and ileum is easy. The contents of the distended cecum are gently "milked" into its lateral recess that is to be resected. The GIA instrument is then used in as many applications as are necessary to resect the cecal cul-de-sac and appendix transversely below the ileocecal valve and the excess lateral cecal pouch with successive vertical placements of the instrument.

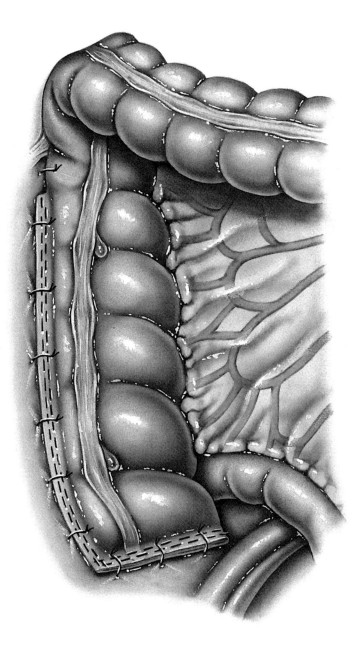

Figure V-18,C: The placement of double rows of staples on each side of the cutting blade prevent any spillage of cecal contents during this operation. The trimmed, normal-appearing cecum, and terminal ileum are now anchored to the retroperitoneal area with interrupted sutures, taking care to protect ureter and iliac vessels.

CURE OF SIGMOID VOLVULUS

The clinical presentation of a sigmoid volvulus is usually more dramatic, since the frequently 360-degrees-plus twist of the omega-shaped sigmoid bowel around its base, supported by a narrow, elongated mesosigmoid results in complete bowel obstruction as well as immediate compromise of the sigmoid vessels. This represents a very early threat to bowel survival and to the patient's life, especially since the condition appears more often in the elderly living in extended care facilities.

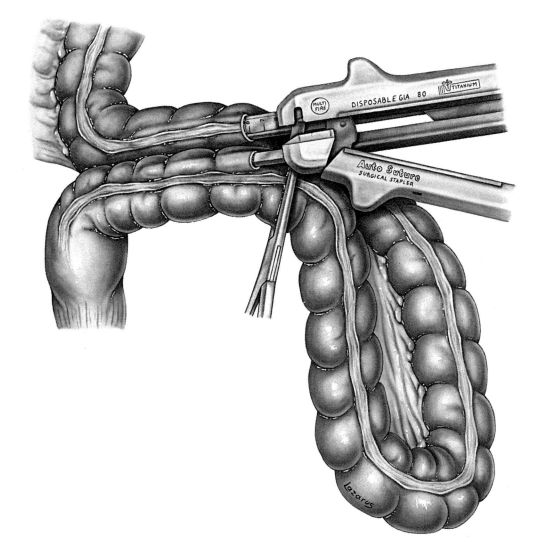

Figure V-19,A: The volvulus is often recognized because of its specific clinical presentation and classical plain radiographic findings. It can therefore be approached through a left McBurney incision. After the affected sigmoid loop has been exteriorized, the accumulated bowel contents are "milked" into the compromised omega loop and clamps are applied at the lines of demarcation and across the narrow mesosigmoid and sigmoid vessels. The loop is then derotated.

The underlying cause, e.g., the narrow mesosigmoid at the base of the omega sigmoid loop, facilitates the use of an "anastomose-resection integree" (anastomosis first - resection second) as shown in Figure V-4,A to F. The GIA instrument is placed through stab wounds into clearly healthy, viable afferent and efferent bowel and is activated, creating a side-to-side anastomosis.

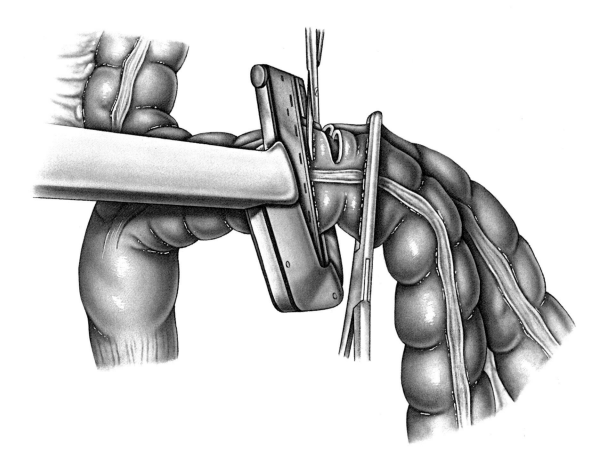

Figure V-19,B: After removal of the GIA instrument, the afferent and efferent bowel loops are clamped and staple closed with the linear TA stapler, in viable areas, just central to the GIA instrument introduction sites. The compromised bowel, including the GIA stab wounds, is excised between the previously applied clamp(s) peripherally and the TA instrument centrally, by using the edge of the TA instrument as a convenient guide for the scissors.

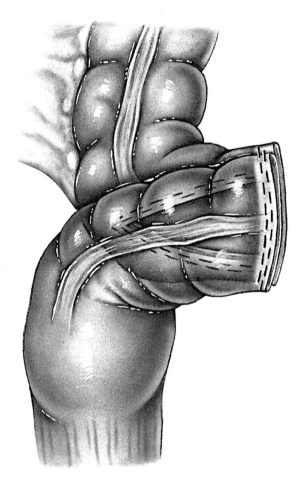

Figure V-19,C: The final result is a widely patent anastomosis in viable bowel. If there is any concern about the immediate survival of this anastomosis, a vented colocolostomy as shown in Figure V-17 represents a nice, safe, temporary alternative, especially in the elderly patient.

LOW ANTERIOR RECTOSIGMOID RESECTION AND TRANSANAL END-TO-END COLORECTAL ANASTOMOSIS

With the arrival of the EEA circular anastomosing instrument in 1977, the true end-to-end, inverting anastomosis with mechanical sutures entered the realm of clinical practice and the attitudes toward staples in surgical operations changed; in part because the anastomotic architecture closely resembled the one of the handsewn end-to-end anastomosis and did not require a major departure from past experience and current practice, albeit at the price of an additional incision and closure of the bowel to place the EEA instrument. This acceptance was, in part, also driven by the realization that anastomoses could be safely performed at 3 to 4 cm from the anal verge, significantly reducing the need for abdominoperineal resections without increasing the incidence of anastomotic failures or strictures; in fact diminishing anastomosis-related complications.

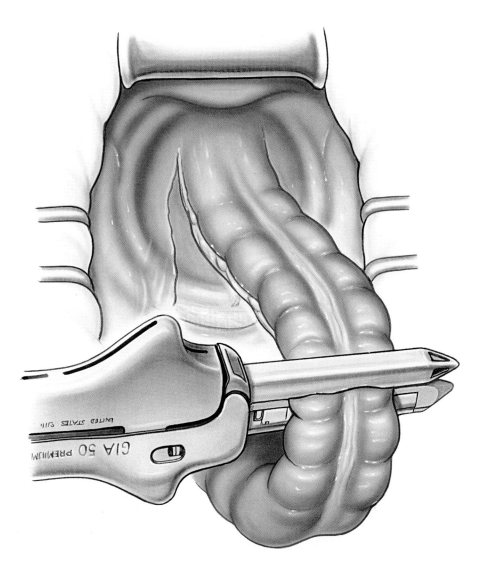

Figure V-20,A: Following mobilization of the sigmoid and descending colon, including the splenic flexure if necessary to gain sufficient bowel length for a tension-free anastomosis later in the pelvis, the sigmoid colon is transected and closed on both sides of this transection with the GIA instrument, at the proximal level chosen for resection. The corresponding vessels can be ligated before this mobilization by following the "no-touch" technique; or in the case of a benign lesion, may be cared for now.

Closure of the colon significantly reduces the risk of contamination and enables instrument-free, gentle traction on the specimen held with a pad, to facilitate further mobilization and dissection into the pelvis, as well as total mesorectal excision, along the "holy plane" of Heald.

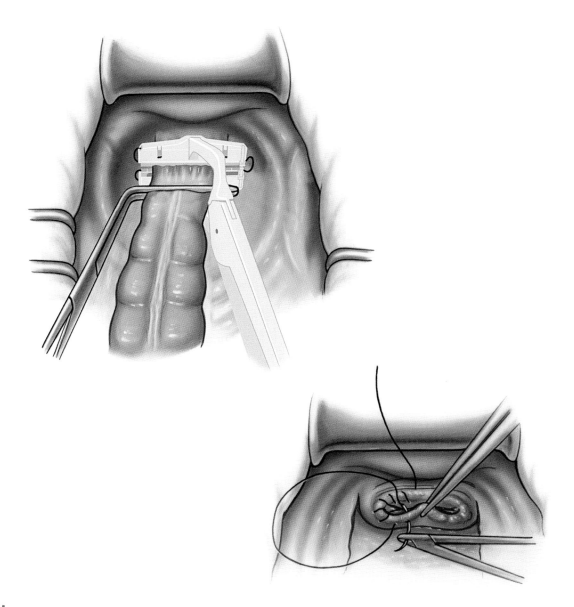

Figure V-20,B,C: The rectum is fitted with a purse-string suture placed either with the Purstring instrument or by a manual whipstitch around the open rectal stump. If the special instrument is used, the rectum is transected along its upper edge and below a right angle clamp. In a case where the purse-string suture is placed on the open rectal rim, the specimen would have been resected below a right angle clamp. Upward pressure on the perineum by an assistant's closed fist will facilitate the placement of the purse-string suture in patients in whom a low anterior anastomosis was not expected and the perineum was not prepared for a separate perineal approach (see Figure V-31, A to E).

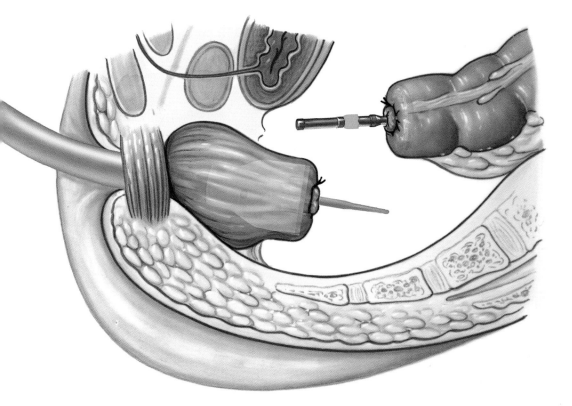

Figure V-20,D: The Purstring instrument is used around the proximal colon just cephalad to the GIA staple line, the purse-string suture is placed, and the staple line excised, caudad to the instrument. The colon caliber is measured to determine the appropriate EEA instrument. The anvil is separated from the instrument and inserted into the colon, and the purse-string suture is tied around the anvil shaft. The EEA instrument is then passed transanally to just below the level of the rectal purse-string suture or the Purstring instrument which is then removed. The center rod is advanced beyond and above this level, and the purse-string suture is tied securely but not too tightly around it, allowing the tissue to slide as the anvil is closed against the cartridge.

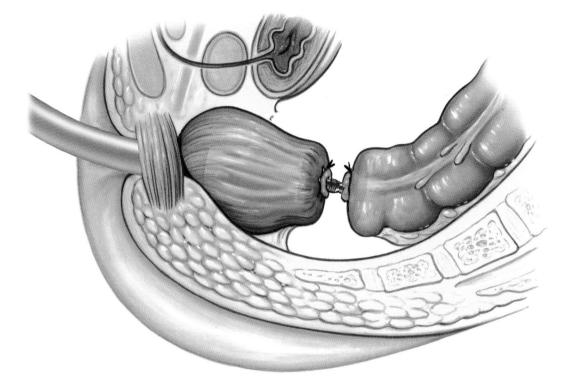

Figure V-20,E: The anvil shaft and center rod are joined. As the anvil is closed against the cartridge, the structures surrounding the anastomotic site are held to ensure that no extraneous tissue is drawn between the cartridge and anvil and thus incorporated into the anastomosis. In addition, the proximal bowel is kept on a gentle stretch superiorly and the EEA cartridge is cautiously pushed against its purse-string from below, so as to create some resistance to the closing movement of the EEA instrument and prevent bunching of the bowel walls at the anastomotic site.

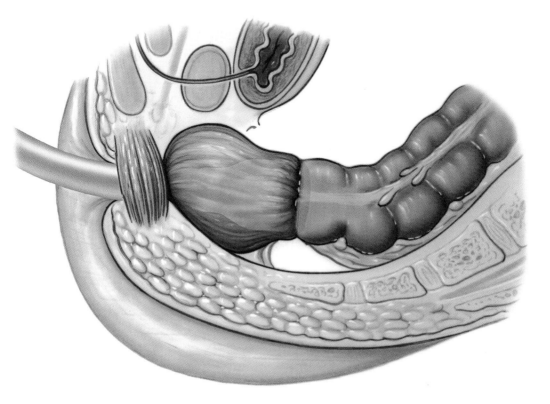

Figure V-20,F: After the anvil has been closed against the cartridge, the handles are activated and the staples driven into the tissues, forming the classical B-shape and securing the end-to-end colorectal anastomosis.

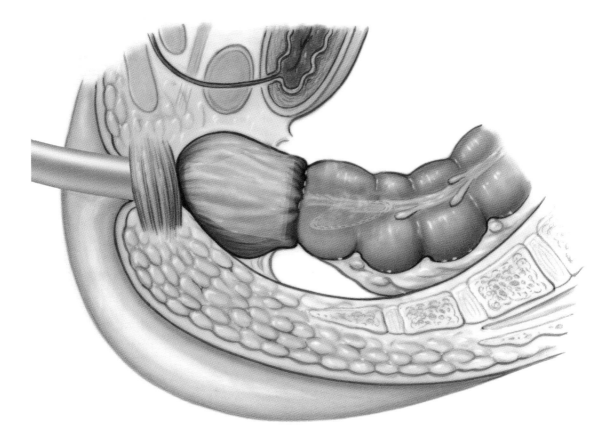

Figure V-20,G: As the anvil is separated from the cartridge in order to disengage the instrument from the anastomosis, the tilt-top anvil configuration facilitates easy removal of the instrument through the anal canal. In the event that there is resistance to removal, a suture can be placed around the free anterior anastomotic rim to lift up the anterior wall of the anastomosis so as to facilitate extraction of the instrument. The excised tissue "donuts" remaining within the instrument are checked to ensure that all tissue layers and purse-string sutures are intact. Depending on the level of the anastomosis, hemostasis can be ensured by direct visualization or with the use of an endoscope placed to, but not through, the anastomotic site.

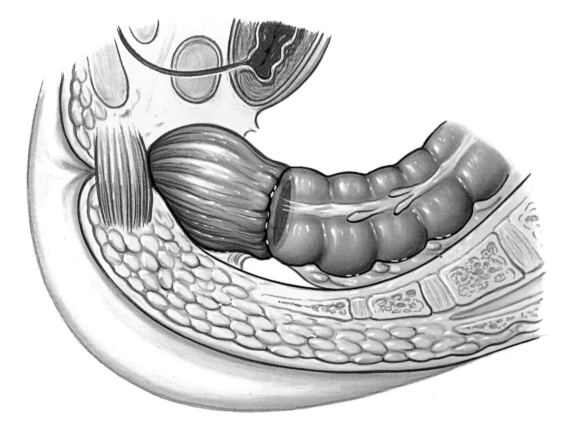

Figure V-20,H: The resulting inverted circular anastomosis is widely patent, and the B-shaped staples ensure rapid healing and an early return of function.

LOW ANTERIOR RECTOSIGMOID RESECTION WITH DOUBLE STAPLING TECHNIQUE FOR COLORECTAL ANASTOMOSIS THROUGH A TRANSANAL APPROACH

Early in our experience with mechanical sutures we had accepted the assumption that it would be dangerous for linear and circular staple lines to intersect without ever exploring this hypothesis experimentally. At the time there was already ample experimental and clinical proof that overlapping staple lines, such as in the functional end-to-end anastomosis, would heal very securely, in contrast to the use of manual sutures where such practices always exposed surgeon and patient to the risk of ischemic angles and corners.

Knight and Griffen put the unspoken ban against intersecting staple lines to the test in 1980 and proved that their "double" stapling technique did not only facilitate very low anterior anastomoses, but made them safer. They did not observe any staple line or circle disruption, nor leaks after underwater testing, during the procedure. Radiographic examination of the intersecting staple lines and circles showed the staples that were in the path of the circular blade to be either transected or pushed aside, without any significant tissue injury that would have been noted at the time of operation. We had observed similar findings with the use of overlapping and intersecting linear GIA staple lines in wedge- and pie-shaped biopsies of the lung, where any tissue damage would have resulted in an immediate air leak.

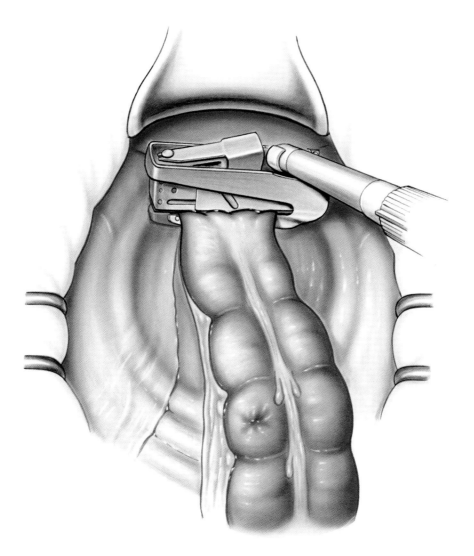

Figure V-21,A: Following mobilization of the future specimen, the rectum is stapled on the specimen side below the tumor and the anorectal canal is irrigated to clear it of debris and intraluminal tumor cells that might jeopardize the short- and long-term health of the anastomosis.

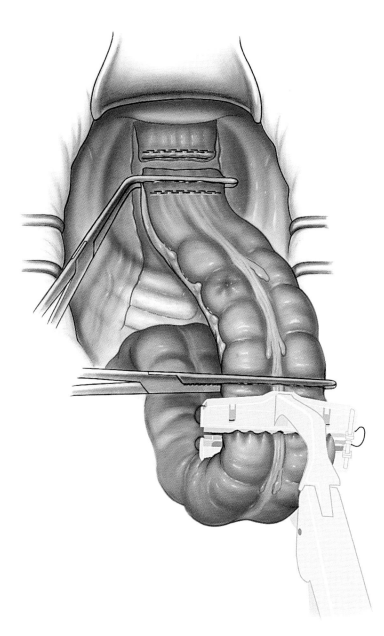

Figure V-21,B: The rectum is then closed at the elected site of resection using the Roticulator instrument that is easier to place, particularly deep in the male pelvis. The handles of the instrument are compressed and a double row of linear staples are placed. With the Roticulator instrument in place, a right angle clamp is applied to the specimen side and the lower rectum is transected using the superior edge of the stapler as a guide. The instrument is removed and the linear closure of the rectal stump is inspected. Proximally, the sigmoid colon can be prepared directly at the site of the elected resection with the Purstring instrument and transected along the instrument edge above a distal clamp.

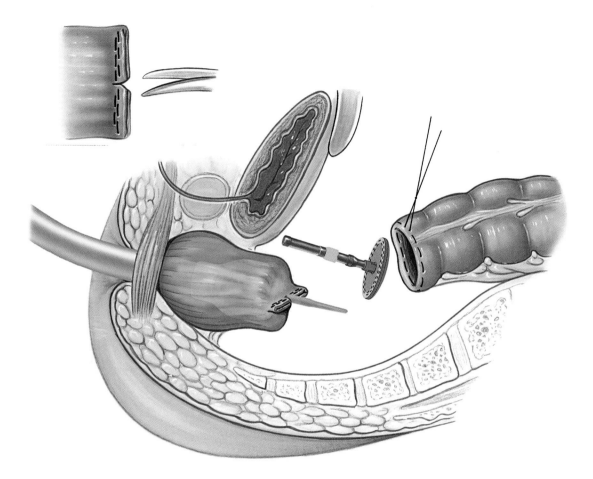

Figure V-21,C: Since the caliber of the anastomosis depends on the diameter of the proximal colon, the luminal cross-section is measured to select the appropriate EEA instrument. The anvil is separated from the instrument and inserted into the proximal colon. The purse-string suture is tied around the anvil shaft. Next, the cartridge-carrying instrument is inserted transanally and gently advanced against the linear rectal closure. A small incision at the midpoint of this closure aids the passage and advance of the center rod of the instrument.

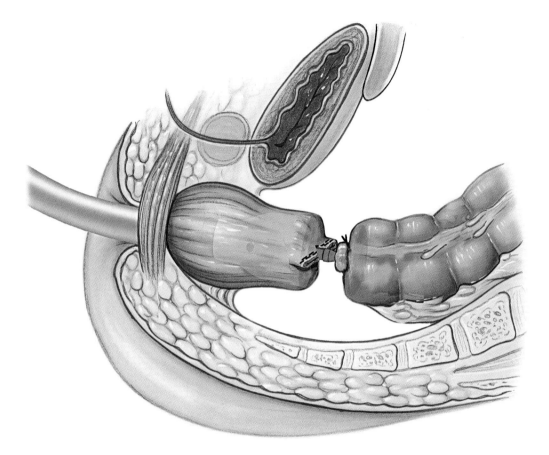

Figure V-21,D: The anvil shaft and center rod are joined. The anvil is closed against the cartridge, taking care to catch no extraneous tissue in this closure, while holding the bowel ends snug against the cartridge and anvil to prevent bunching or pleating of tissue within the anastomosis.

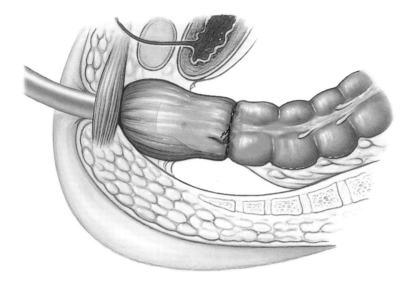

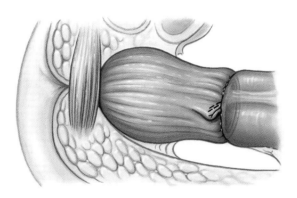

Figure V-21,E: The instrument is activated placing a double staggered circle of staples joining rectum to colon, while the circular blade cuts through the colon and stapled rectal stump creating the stoma.

Figure V-21,F: Following instrument removal, the excised tissue "donuts" remaining within the instrument are examined to ensure that all tissue layers are present within the intact donuts. The anastomosis is checked for hemostasis and competency, with special emphasis on the crossings of circular and linear staple lines.

LOW ANTERIOR RECTOSIGMOID RESECTION WITH TRIPLE STAPLING TECHNIQUE FOR COLORECTAL ANASTOMOSIS THROUGH A TRANSANAL APPROACH

In assessing the factors that made the double stapling technique not only possible, but produced a better and simpler anastomosis, the old dictum that the combination of motivation and creative research to improve an accepted gadget can always lead to a better mousetrap came to mind. The discrepancy in caliber and tissue consistency between the wider and muscular rectum and the narrower and thinner proximal colon appeared as one factor that could adversely influence the end-to-end joining of these hollow viscera, especially since the caliber of the joining instrument is dictated by the diameter of the narrow proximal colon. Furthermore, the incarceration by purse-strings of bowel rims that are different in circumference and tissue consistency can conceivably facilitate the evasion of a pleat or two of the bowel, tied around anvil shaft and center rod, especially as the tissues are compressed between anvil closing onto cartridge, leaving scant space for a fleshy rectum for instance. In our experience this would always occur on the posterior aspect of a deep anterior anastomosis, making repair irritatingly difficult.

Knight and Griffen's anastomosis eliminated the discrepancy of size and tissue consistency between rectum and colon as a complicating factor by creating a flat rectal surface with the linear closure, of which only enough cross section would be involved to satisfy the circumference of the proximal colon, carried by an appropriately sized anvil.

As a consequence of this reasoning, the German proverb "doppelt genäht hält besser" led us to reason that the absence of a purse-string could be beneficial not only on the rectum, but also on the colon, especially since the transverse rectal linear closure and the sagittal colon closure can be superposed in a cross-like fashion. This will require staple line transections, one only at each site: 3 o'clock and 9 o'clock for the rectum, 6 o'clock and 12 o'clock for the colon.

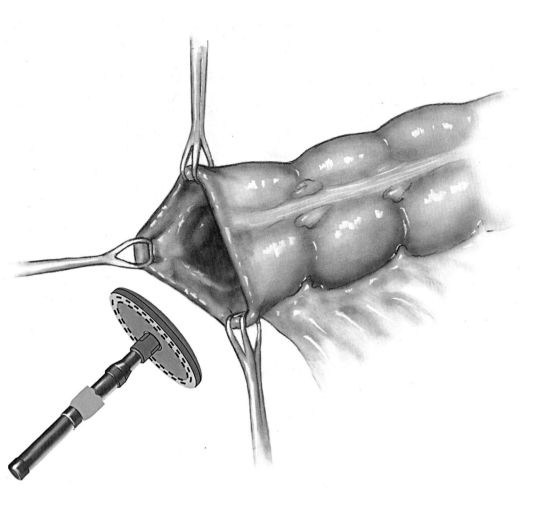

Figure V-22,A: Following rectosigmoid resection, the lumen of the proximal colon is sized and the appropriate anvil is inserted into the open proximal bowel.

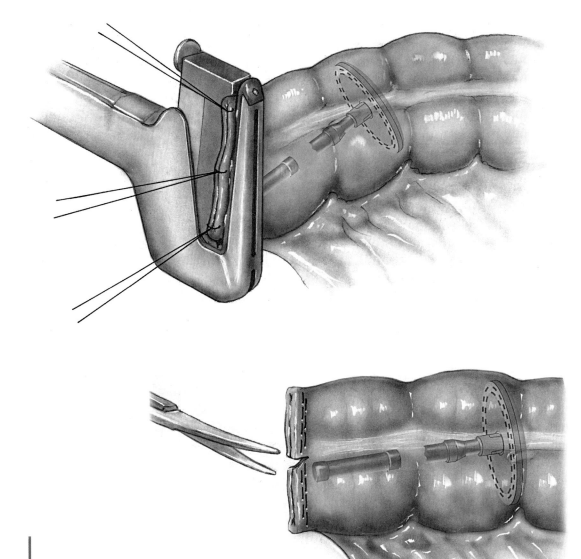

Figure V-22,B: The proximal colon containing the anvil and its shaft is closed with a TA instrument in the same plane as the mesocolon.

Figure V-22,C: A small incision is made at the midpoint of the stapled closure to aid in retrieving the anvil shaft.

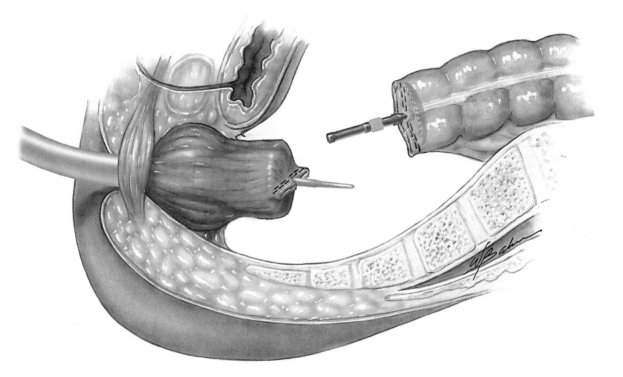

Figure V-22,D: The EEA instrument is inserted transanally, and the cartridge is advanced to the level of the rectal closure and held against it from below. The tip of the center rod is advanced to perforate the rectal staple line at its midpoint. The anvil shaft and center rod of the instrument are engaged.

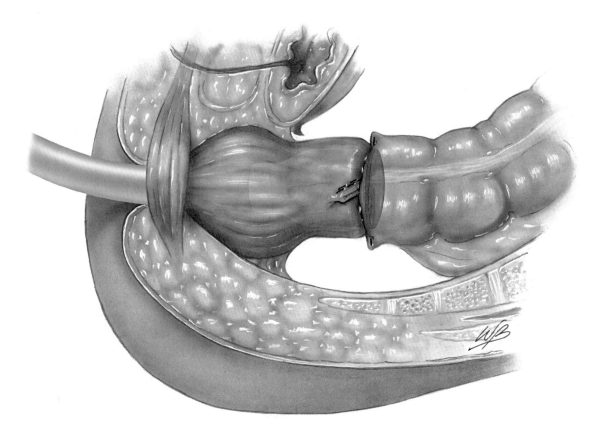

Figure V-22,E: The anvil is closed against the cartridge, taking care not to incorporate any extraneous tissue into this closure, while keeping the bowel ends snug against the cartridge and anvil to prevent bunching. The colon is positioned so that the two linear staple lines are at right angles to one another in a cross-like mode, as described above. The instrument is activated placing a circular double staggered row of anastomosing staples, and the knife blade cuts the stoma through the linear staple lines at four points.

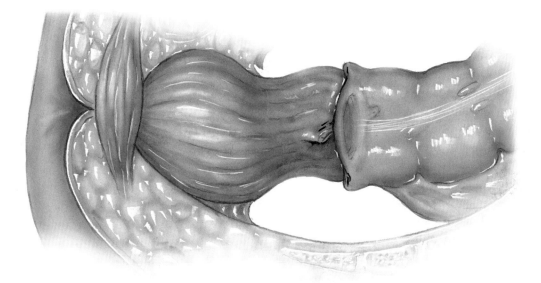

Figure V-22,F: Following instrument removal, the excised tissue "donuts" remaining within the instrument are examined to ensure that all circular tissue layers are intact and the anastomosis is checked for competence and hemostasis. The four points where the circular and linear staple lines cross are inspected and the rectal dog ears at 3 o'clock and 9 o'clock can be attached to the lateral colon wall, as the 6 o'clock and 12 o'clock colon dog ears can be sutured to the anterior and posterior rectal wall. In the rare instance where both the colon and rectum are wider than the largest circular anastomosing instrument, the triple stapling technique represents a safe solution to a technical challenge. Together with the anterior perineal approach to low colorectal reconstruction (Figure V-31, A to E), it has become our procedure of choice in the very low anterior resection.

ANTERIOR RECTOSIGMOID RESECTION AND TRANSABDOMINAL END-TO-END COLORECTAL ANASTOMOSIS

If the use of mechanical sutures limited the technical demands of bowel reconstruction to a few, rigid procedures, stapling would obviously fail to solve the variety of requirements and surprises that present themselves during a given operation. The need for flexibility and adaptability is especially obvious at both extremes of the alimentary tract.

In patients scheduled to undergo a "simple" sigmoid resection, especially for benign or premalignant disease, it becomes often necessary to extend the dissection distally, moving the anastomosis to mid-rectum, when transanal placement of the EEA instrument was not anticipated or in fact is not necessary.

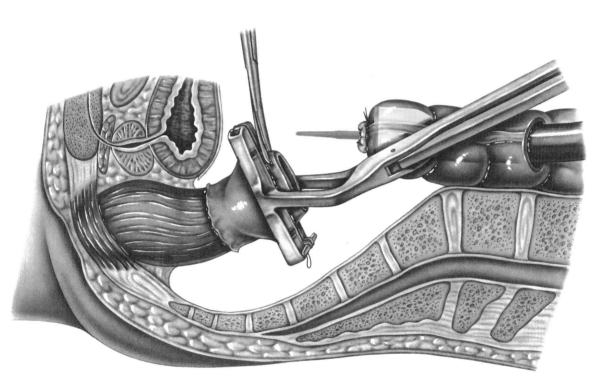

Figure V-23,A: Following rectosigmoid resection, the proximal sigmoid colon and the upper or mid-rectum are prepared for a traditional end-to-end stapled anastomosis, by placing purse-string sutures at both bowel ends. A colotomy is performed approximately 4 to 5 cm proximal to the future anastomotic site, and the appropriately sized EEA instrument, without anvil, is inserted to the level of the purse-string suture on the colon. The center rod is fully advanced beyond this level and the purse-string suture is tied snugly but not too tightly around it, allowing the tissue to slide as the anvil is closed against the cartridge.

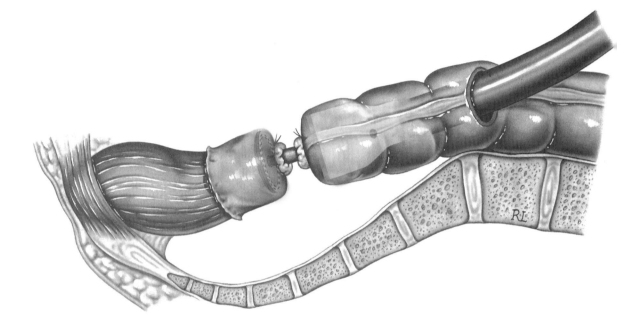

Figure V-23,B: The anvil is inserted into the rectum and the purse-string suture is tied securely around its shaft. The anvil shaft and center rod are mated, and the instrument is closed and activated, placing a double circle of staples and transecting both purse-stringed bowel cul-de-sacs with the circular blade, inside the staple circles.

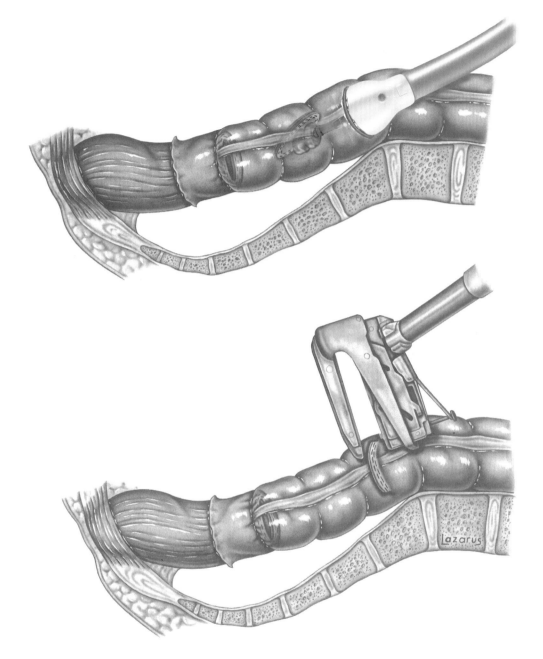

Figure V-23,C: The anvil and cartridge are disengaged from each other and the instrument is removed. The excised tissue "donuts" remaining within the instrument are examined to ensure that all tissue layers are represented within the circular bowel ends gathered by the purse-string sutures.

Figure V-23,D: Following inspection of the anastomosis, the colotomy is closed transversely with a linear stapler in an everting fashion.

ANTERIOR RECTOSIGMOID RESECTION AND TRANSABDOMINAL SIDE-TO-END COLORECTAL ANASTOMOSIS

A result similar to the anastomosis shown in Figure V-23 can be obtained by using the open, proximal colon for the placement of the EEA instrument, instead of the separate proximal colotomy, to achieve a side-to-end colorectal anastomosis. This version creates a golf club configuration of the colon proximal to the anastomosis that functions as a small pouch and has been beneficial in the early postoperative regulation of bowel movements (Adloff, 1980).

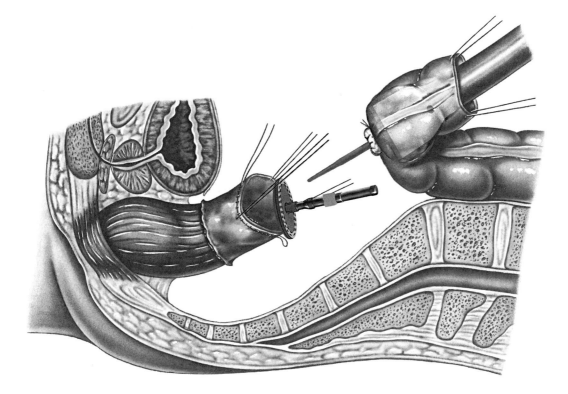

Figure V-24,A: The EEA instrument is inserted into the open proximal colon and turned so as to place the center rod of the instrument against the antimesocolic border of the bowel at the site of the proposed anastomosis. The center rod is advanced, allowing its tip to protrude through the wall of the colon. A manual purse-string suture is placed around the perforation to prevent any serosal tearing or tissue slippage as the instrument is closed. This suture is tied snugly but not too tightly around the center rod of the instrument to prevent invaginating tissue within the cartridge as it is closed against the anvil. The rectum is prepared with a purse-string suture and, with the aid of three traction sutures or Allis clamps keeping the lumen open, the anvil is inserted.

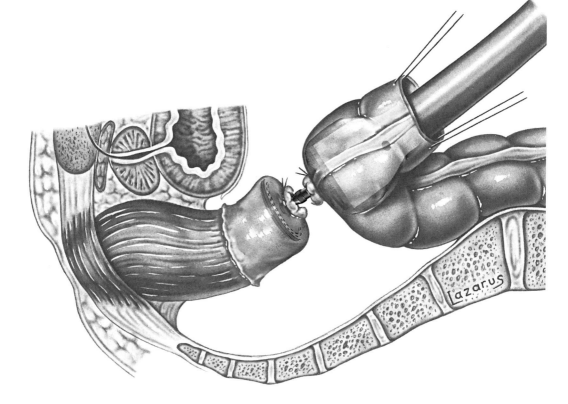

Figure V-24,B: The purse-string suture is tied around the anvil shaft, and the center rod and anvil shaft are joined. As the anvil and cartridge are closed against each other, care is taken not to incorporate any tissue folds at the level of the proximal colon angulation.

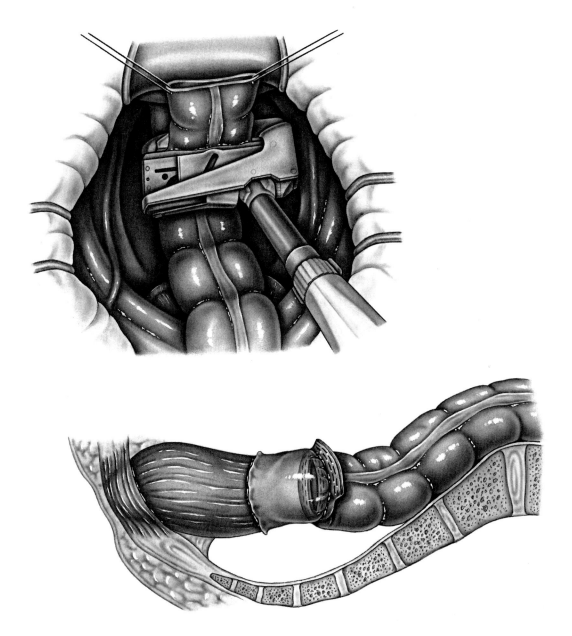

Figure V-24,C,D: After creating the anastomosis, the proximal colon is closed with a linear stapler and the redundant tissue is excised, leaving the previously described golf club configuration of the proximal colon.

ANTERIOR RECTOSIGMOID RESECTION AND TRANSABDOMINAL SIDE-TO-SIDE COLORECTAL ANASTOMOSIS

While the circular end-to-end stapled anastomosis has gained wide acceptance almost from the start (1977), the side-to-side anastomosis in a variety of incarnations remains a real alternative, for more than the historical reasons of having been available first (1968). Regardless of the suture materials used, all other factors being equal, the side-to-side anastomosis is recognized to be structurally the strongest because it is the least prone to technical flaws at the time of construction and is the most vigorous in its short- and long-term ability to heal. Added to that is the fact that the GIA instrument, easier to use than the EEA instrument, presents a built-in mechanical simplicity that sidesteps most of the potential failings of the tedious, stitch-by-stitch, manual sutures. Because of the need to have minimal lengths of bowel available on both sides of the anastomosis, the side-to-side technique is best suited for joining bowel to the mid- or upper rectum.

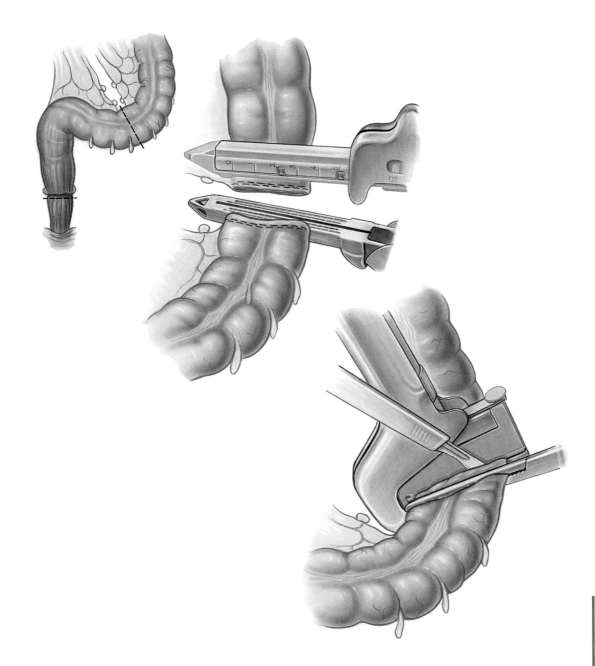

Figure V-25,A,B,C: The segment to be resected is mobilized, including the corresponding mesocolon, lymph nodes, and supporting blood vessels, ligated high at their respective origins. The proximal colon is closed on both sides of the transection with the GIA instrument or with the TA instrument on the patient's side, in which case the bowel is transected between the linear stapler and a Kocher clamp on the specimen side. The rectum is severed distal to a clamp that occludes the specimen, which is removed from the operative field. The pelvis is packed around the open rectum which is held open with stay sutures and gently irrigated and cleared by aspiration of any debris. The packing is removed.

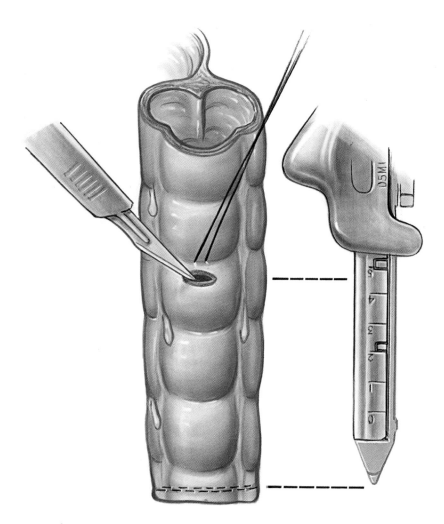

Figure V-25,D: The sigmoid colon is positioned so that the mesocolon will come off posteriorly, opposite the anterior anastomotic site. The sigmoid cul-de-sac is measured starting at the level of its closure and going proximally, over the length that corresponds to the fork of the GIA instrument used to perform the anastomosis. An anterior antimesocolic stab wound is placed at this proximal level, guarded by a stay suture to mark the spot and help with the placement of the GIA anvil.

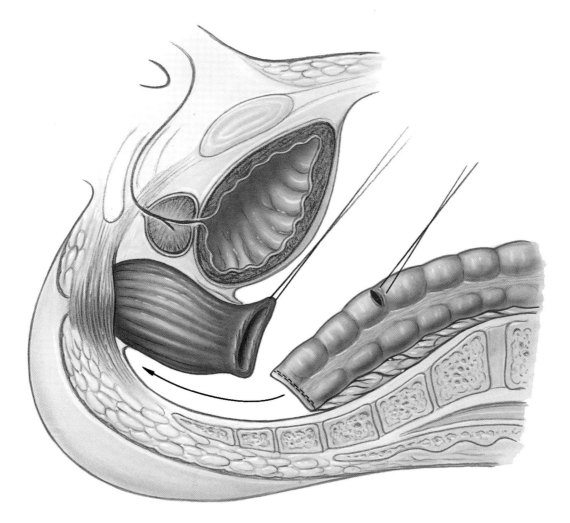

Figure V-25,E: The posterior wall of the rectum is cleared and the proximal colon is positioned into the retrorectal space.

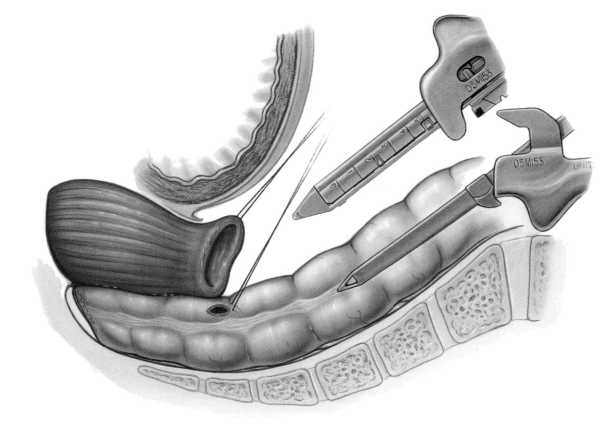

Figure V-25,F: With the aid of the previously placed traction (stay) sutures, the anvil fork of the GIA instrument is advanced into the sigmoid colon along its antimesocolic border and the cartridge fork is positioned into the open rectum. To ensure maximum stomal size, insert the forks fully into each lumen.

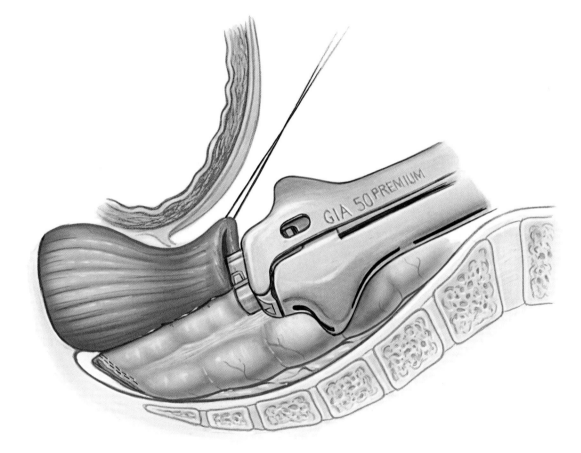

Figure V-25,G: The instrument halves are joined, and the instrument is locked and activated, creating the side-to-side anastomosis.

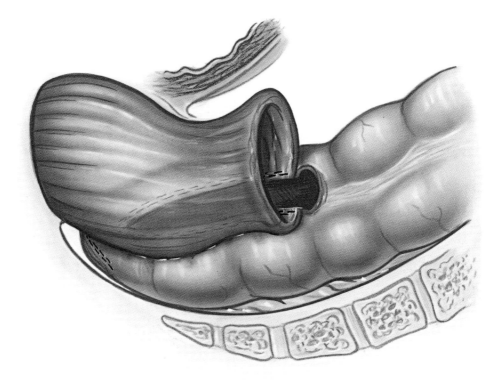

Figure V-25,H: Following removal of the instrument, the anastomosis is examined to ensure hemostasis.

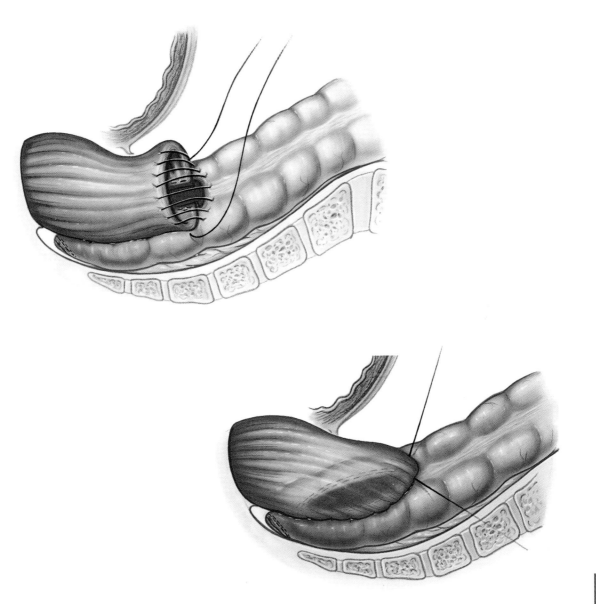

Figure V-25,I,J: The now common opening is closed manually. The tight confines of the distal pelvis protect the small sigmoid cul-de-sac from distending.

ANTERIOR RECTOSIGMOID RESECTION AND TRANSABDOMINAL SIDE-TO-SIDE COLORECTAL ANASTOMOSIS WITH "ANASTOMOSE-RESECTION INTEGREE" ANASTOMOSIS FIRST - RESECTION SECOND

This technique is most helpful in the patient with a narrow pelvis in whom gentle traction on the specimen, kept in continuity with the rectum, facilitates the side-to-side anastomosis, especially at the level of the mid-rectum. It represents a modification of the functional end-to-end anastomosis that makes stapling in narrow spaces possible.

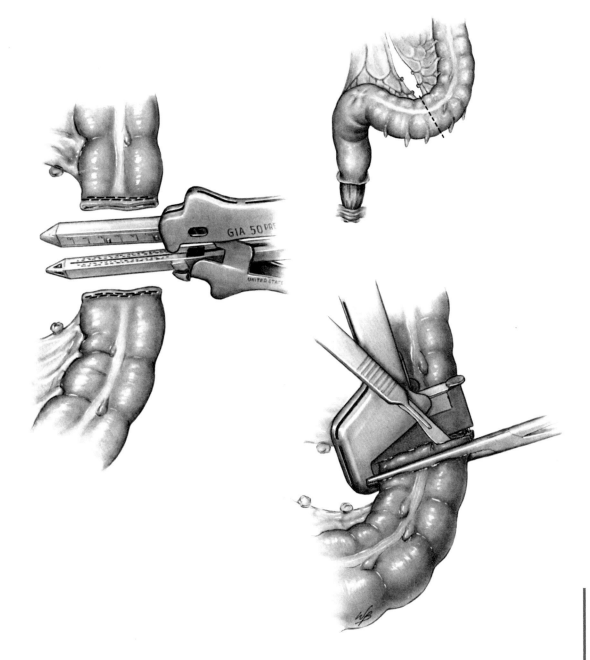

Figure V-26,A,B,C: The segment to be resected is identified and mobilized, together with corresponding mesocolon, lymph nodes, and supporting blood vessels, ligated at their origins. The proximal colon is closed with either the GIA or TA instrument, as shown and described in Figure V-25, A to C.

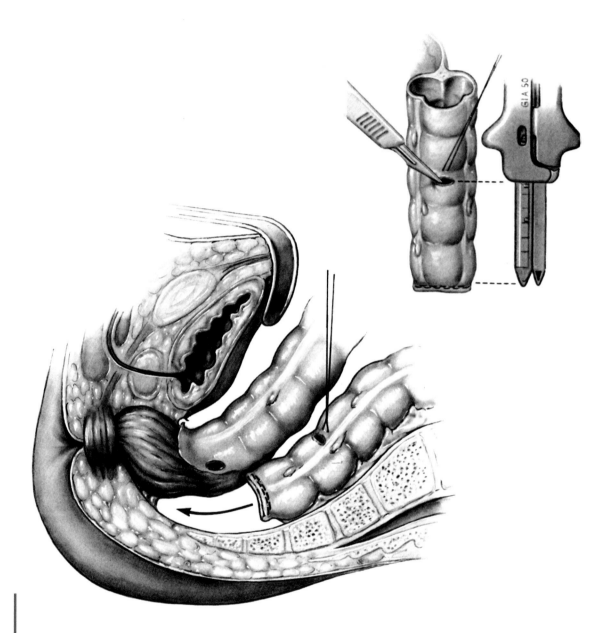

Figure V-26,D: The proximal sigmoid colon is positioned so that the mesosigmoid will come off posteriorly and the site for anastomosis will be on the anterior, antimesocolic wall. The sigmoid cul-de-sac is measured to correspond to the length of the fork of the GIA instrument used to perform the anastomosis and a suture is placed at the site of the stab wound created for the introduction of the anvil-bearing fork.

Figure V-26,E: A corresponding stab wound is made in the left anterolateral wall of the rectum at the proximal point of the proposed anastomosis. The posterior wall of the rectum is dissected free, and the colon is positioned in the retrorectal space.

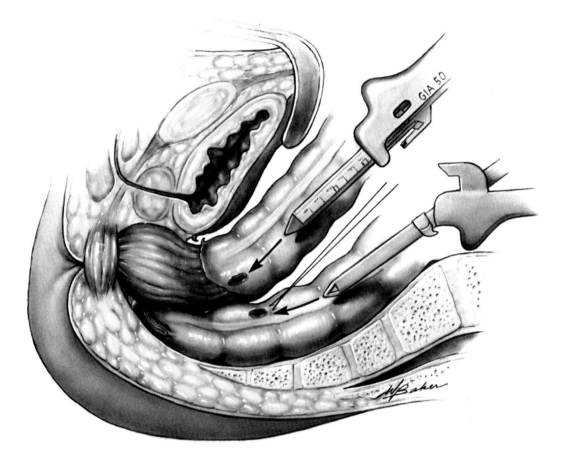

Figure V-26,F: With traction on the specimen to elevate the rectum, the anvil fork of the GIA instrument is placed into the lumen of the sigmoid colon, along the antimesenteric border, and the cartridge fork is advanced into the rectum along the left lateral rectal wall.

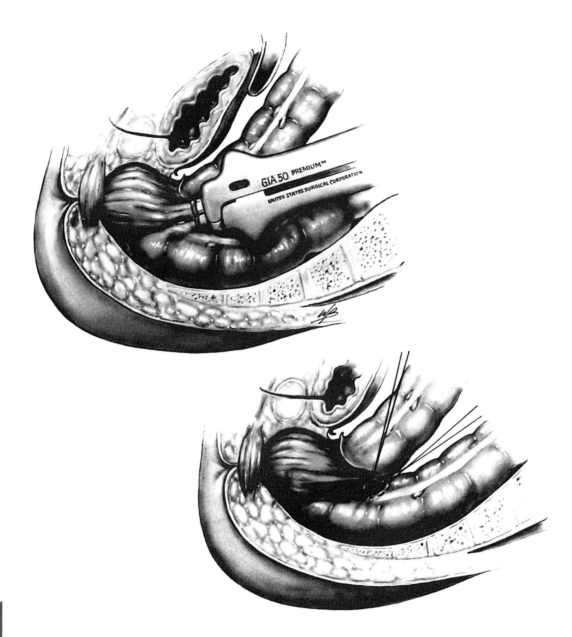

Figure V-26,G: The instrument halves are joined and the instrument is locked and activated, creating the side-to-side colorectal anastomosis.

Figure V-26,H: An everting stay suture is placed around the end of each anastomotic staple line, to be held in opposite directions, so as to spread the now common opening of the GIA introduction site transversely and facilitate the closure of this opening, together with and against the anterior rectal wall.

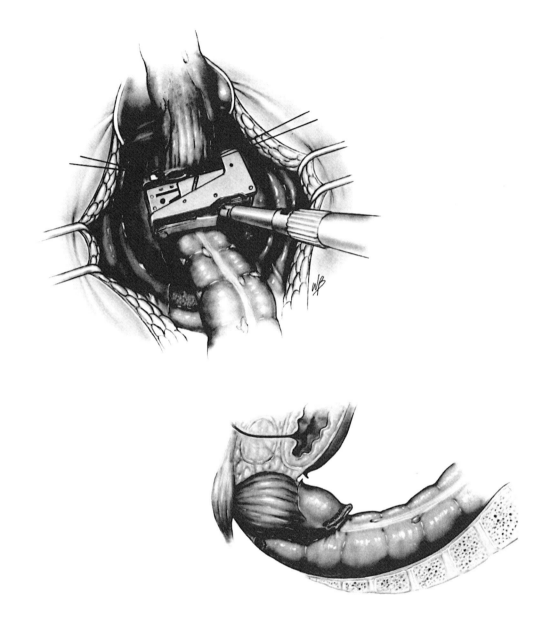

Figure V-26,I: With the specimen on gentle traction and held vertically through the lower angle of the incision, behind the pubis, the Roticulator stapling instrument is placed around the rectum and the stay sutures that spread transversely the common GIA opening, to include the ends of the staple lines and the posterior lip of this opening into the closure of both the common opening and the rectum. The instrument is activated and the specimen is resected using the upper edge of the Roticulator instrument as a guide.

Figure V-26,J: The sigmoid cul-de-sac is prevented from distending by the narrow surrounding pelvis at this point.

ANTERIOR RECTOSIGMOID RESECTION AND TRANSABDOMINAL TRUE FUNCTIONAL END-TO-END COLORECTAL ANASTOMOSIS

By reducing the duration and size of open bowel exposed in the peritoneal cavity, this technique is especially suited to protect the patient from intra-abdominal contamination in those situations where the preoperative bowel preparation was not satisfactory. In such a case, the retained large-bowel contents can be "milked" from the rectum upward and from the descending colon downward, into the future specimen.

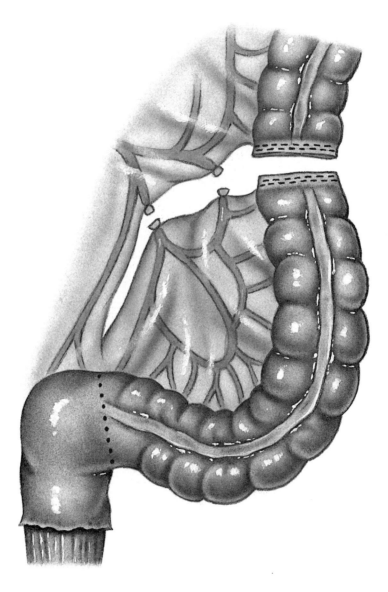

Figure V-27,A: The specimen is delineated as in the previously described procedures, Figure V-25 and Figure V-26. The sigmoid colon and upper rectum are then transected at the selected sites, with two applications of the GIA instrument that close the bowel on both sides of the transection, preventing all and any leak from the bowel. The everted mucosal rims are cleaned with cotton swaps soaked with antiseptic solution.

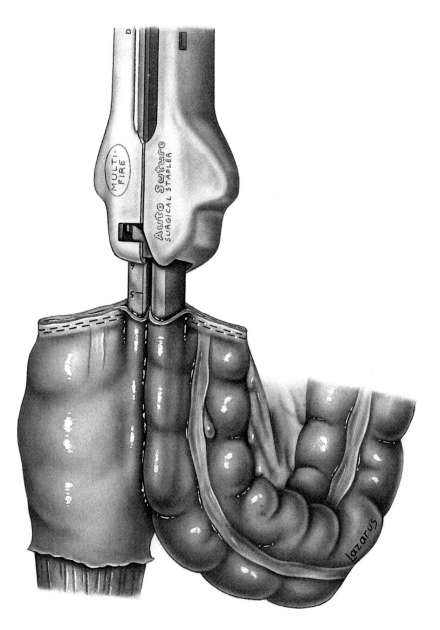

Figure V-27,B: The naturally curved descending colon is placed alongside the lateral aspect of the rectal stump, with the stapled end turned superiorly. The antimesocolic and antimesorectal corners of both bowel ends are excised, the GIA instrument is placed with one fork in each bowel lumen, the instrument halves are matched, locked, and the instrument is activated, achieving a functional end-to-end rectocolostomy.

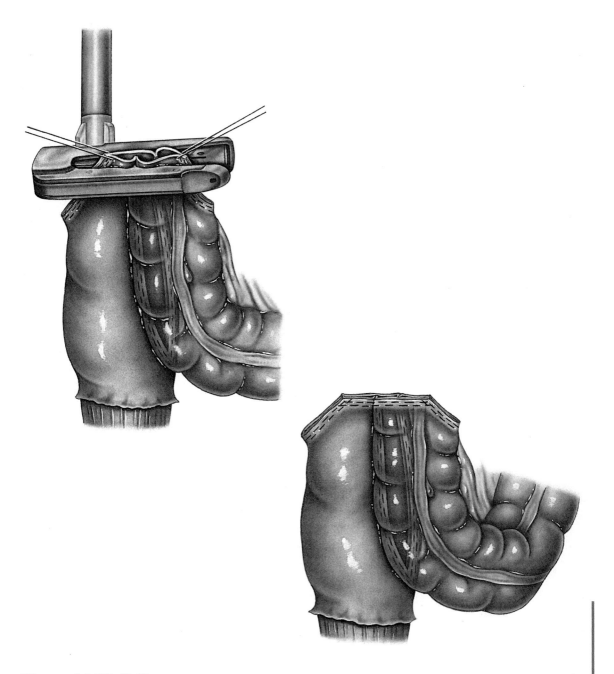

Figure V-27,C,D: After removal of the GIA instrument, the now common entry site is closed with the linear, Roticulator instrument. The efficiency of this anastomosis makes it a favorite when expeditious and tidy action is essential.

ANTERIOR RECTOSIGMOID RESECTION AND TRANSABDOMINAL FUNCTIONAL END-TO-END COLORECTAL ANASTOMOSIS WITH "ANASTOMOSE-RESECTION INTEGREE" ANASTOMOSIS FIRST - RESECTION SECOND

This represents one more variation on a common theme and like all previous side-to-side anastomoses at this level in the pelvis, is performed most comfortably, if it is to the upper rectum. Again by leaving the specimen attached to the rectum and exerting gentle traction upwards, the anastomosis of the sigmoid colon, bent on itself with the staple line facing upwards, is facilitated.

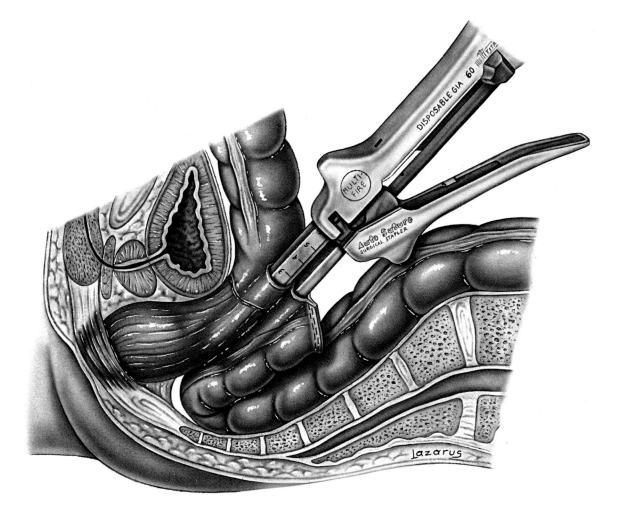

Figure V-28,A: After the specimen with corresponding mesocolon, lymph nodes, and vessels has been liberated, it is transected at the selected proximal level of separation from the remaining bowel and closed on both sides of this transection with the GIA instrument. While maintaining gentle vertical traction on the specimen left in continuity with the rectum, a stab wound is placed into the left lateral wall of the upper rectum. The proximal colon, bent on itself with the staple line facing upwards, is moved into the retrorectal space and the antimesocolic corner of the staple line is excised. The GIA stapler is placed with one fork entering each lumen through the respective stab wound. The instrument halves are matched, locked, and the instrument is activated, creating a side-to-side anastomosis.

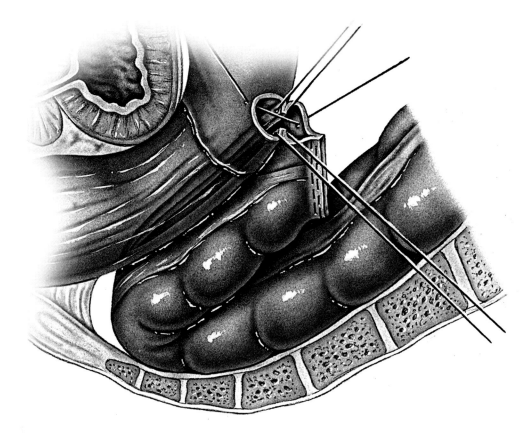

Figure V-28,B: After removal of the GIA instrument, stay sutures are placed, one at each end of the linear, anastomosing staple lines.

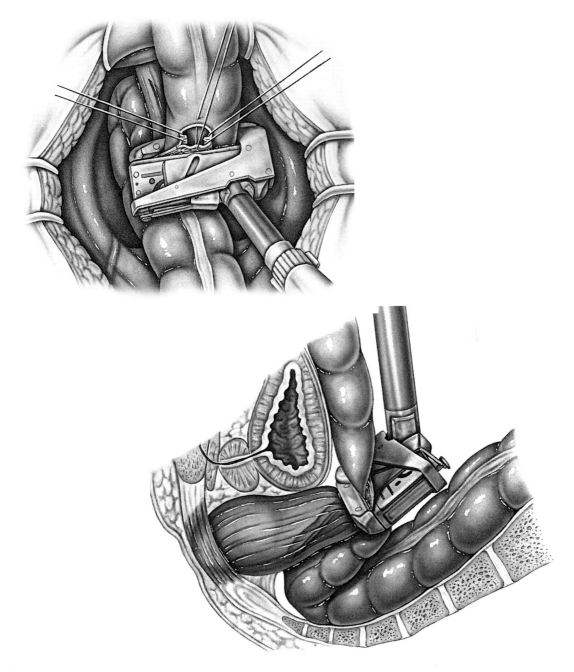

Figure V-28,C: These stay sutures are pulled in opposite directions, so as to approximate the lips of the now common GIA introduction site. An additional suture at midpoint between the two stay sutures aids in the pouting eversion of the rim of the common opening. The Roticulator stapler is placed around the vertically held rectosigmoid junction and deep to the sutures maintaining the everted rim above the closing jaw of the linear stapler.

Figure V-28,D: The instrument is closed and activated, stapling simultaneously the lumina of the rectum and of the common opening.

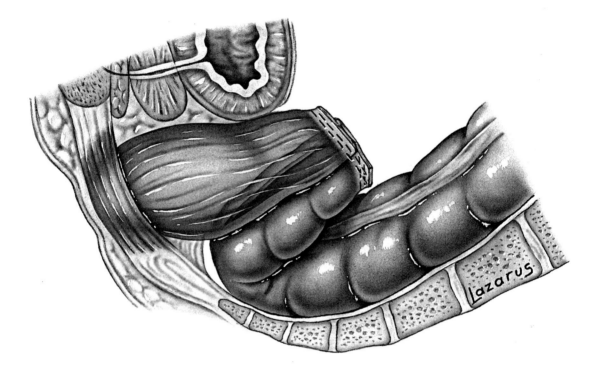

Figure V-28,E: The specimen is excised, using the upper edge of the Roticulator instrument as a guide. The final result is a true functional end-to-end rectosigmoidostomy with wide triangulated patency.

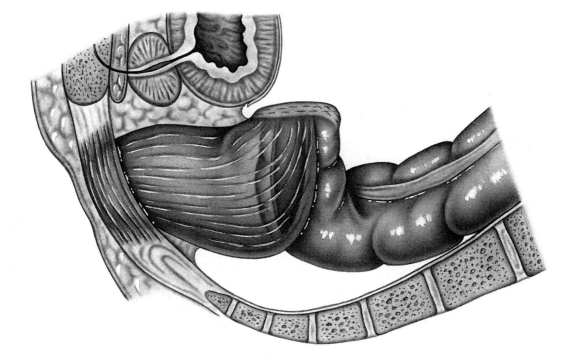

Figure V-28,F: With time, the anastomotic site assumes a straight line, with the original angulation of the proximal limb converting to a golf club configuration and the radiological and endoscopic appearance of an end-to-end anastomosis.

SUBTOTAL COLECTOMY AND CECUM TO RECTUM CIRCULAR END-TO-END ANASTOMOSIS

This operation, proposed by Mouiel and his associates in 1985, saves the ileocecal valve in patients who require a near-total colectomy, usually for a mixture of benign and malignant lesions of the large bowel that present at different levels of implantation, but where the distal rectum and the cecum are always clearly free of disease. Furthermore, the blood supply, always of some concern in the area of the ileocecal valve and ascending colon, does not require any more than the usual attention in all bowel resections and anastomoses, since the continuity of ileum and cecum is not interrupted. While the field of application is admittedly small, patients who have benefited from this procedure are grateful for postoperative bowel habits that are easily manageable: three to four soft bowel movements per 24 hours. Furthermore, the postoperative monitoring of the remaining cecum and rectum is easily accomplished with rigid or flexible endoscopy.

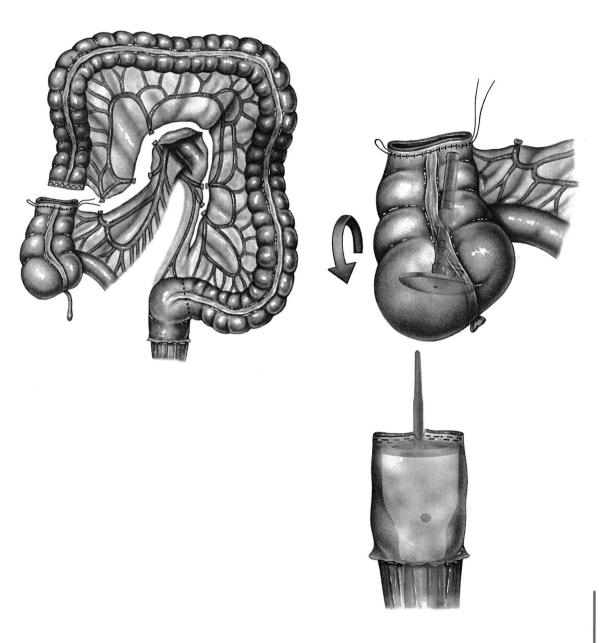

Figure V-29,A: The subtotal colectomy is performed between a purse-string placement on the upper cecum and a linear staple closure of the rectum.

Figure V-29,B: The EEA instrument, without anvil, is inserted transanally to the level of the rectal stump closure. The center rod is advanced to perforate the staple line at the midpoint of the linear rectal closure. The anvil is placed into the open cecum, head down. The cecum is rotated 180 degrees anteriorly, around the long axis of the terminal ileum, without jeopardizing ileal patency or blood supply.

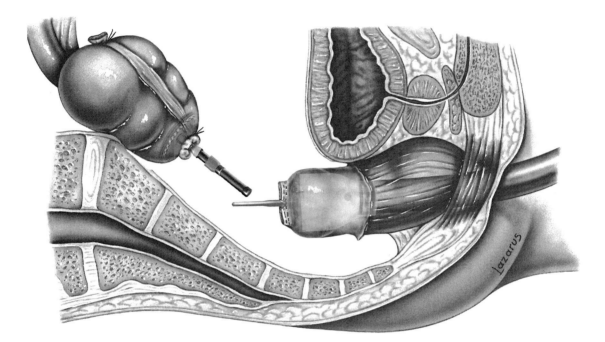

Figure V-29,C: The purse-string suture is tied around the anvil shaft, and the anvil and cartridge are mated and closed against each other. The instrument is activated, placing two staggered circles of staples while the circular blade cuts out the by now familiar "donuts" of rectum and cecum.

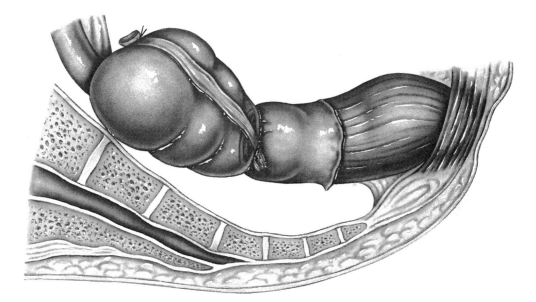

Figure V-29,D: The anastomosis is checked transanally by endoscopy with the abdomen open.

TOTAL COLECTOMY WITH SPHINCTER PRESERVATION, CREATION OF AN ILEAL J-POUCH, AND ILEOANAL ANASTOMOSIS

The development of the continent ileal reservoir by Kock in the mid 1970s represents the first effort to provide a better quality of life to patients who had to undergo total colectomy, usually for complications of ulcerative colitis, including premalignant changes, and for familial polyposis. These conditions, if treated within time limits that allowed their long-term containment, always led to a life with a constantly active ileostomy. The ileal reservoir by Kock eliminated this constraint, although it did not always completely permit the abandonment of the collecting bag. The transfer of the ileal reservoir technology into the pelvis, with ileoanal anastomosis and substitution of Kock's technically complex and functionally often capricious continent nipple by the intact anal sphincters, was a great step forward in the quality of life of these patients, albeit at the price of a period of postoperative readaptation in bowel habits, at times demanding great discipline and understanding of untoward events, especially of and by the younger patients. A variety of ileal pouch configurations was developed, however, at present, the J-pouch appears to have best resisted the tribulations of time and the trials of physical wear and tear.

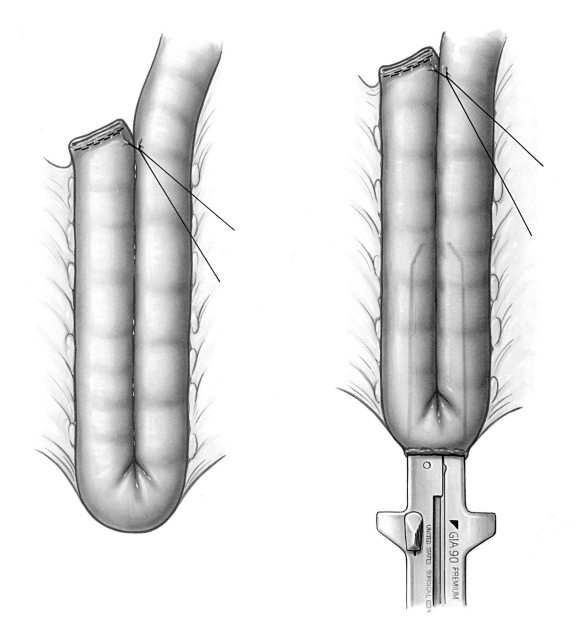

Figure V-30,A: After total colectomy has been performed, leaving a 5 to 6 cm rectal stump, the terminal ileum is closed and transected with the GIA instrument, flush with the cecum. The mesenteric root is freed to allow the future ileal reservoir to reach the anus without tension. The terminal 25 to 36 cm of ileum are looped into the outline of a 12 to 18 cm "J"-shaped pouch. The antimesenteric borders of the J-shaped bowel loop are approximated with a stay suture.

Figure V-30,B: A 2 cm incision is made in the antimesenteric wall of the curved jejunum, exactly at the bend between both limbs of the loop. One fork of the GIA instrument is placed into each limb of the loop, along the antimesenteric walls, and the instrument halves are matched and locked. The instrument is activated to achieve a long side-to-side anastomosis. The GIA instrument halves are disengaged from each other and removed.

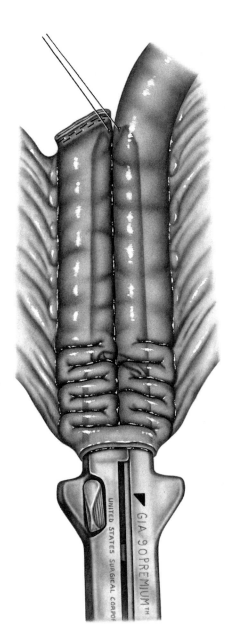

Figure V-30,C: Through the same opening, the procedure is repeated by telescoping the anastomosed ileum downward and advancing the forks into the lumina of the yet intact ileal limbs, while making sure that the superior angle of the previous staple lines and the crotch of the instrument are matched as the instrument is closed to ensure security and continuity of the staple lines along the antimesenteric borders. In general, two applications of the GIA 80 or 90 stapler are required to complete the pouch. Hemostasis is ensured.

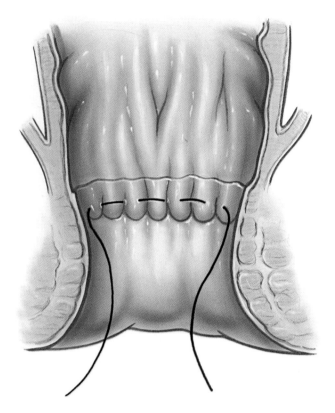

Figure V-30,D: After accomplishing a mucosectomy of the previously left rectal stump, a manual purse-string suture is placed 0.5 cm above the dentate line, into a mucosal collar, just below the level of where the mucosectomy had been stopped, 1 to 2 cm above the dentate line.

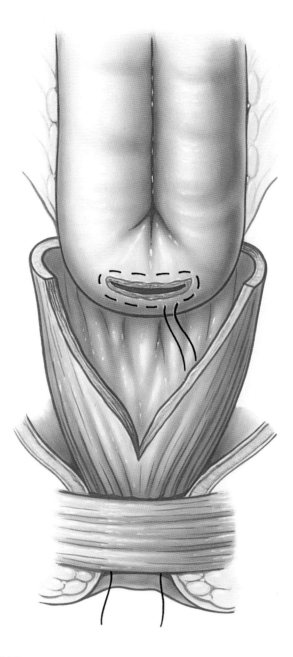

Figure V-30,E: The anterior rectal wall is then incised vertically to create a receptacle that will better accommodate and retain the pouch. A second manual purse-string suture is placed around the ileal GIA placement incision. The anvil of the EEA instrument is positioned into the pouch and secured by tying the suture into the notch on the anvil shaft.

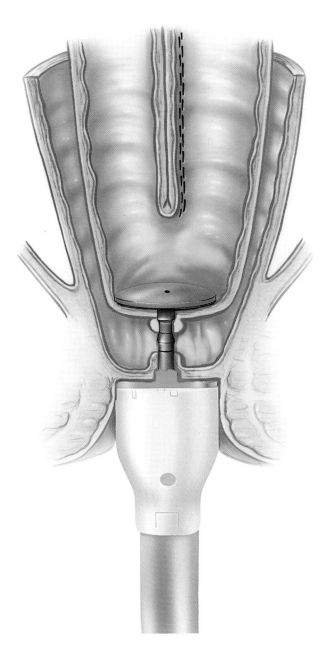

Figure V-30,F: The EEA instrument carrying the cartridge is introduced into the anus and opened to advance the center rod. The anal mucosal purse-string suture is tied around the center rod. The anvil shaft and center rod are joined, and the instrument is closed and activated.

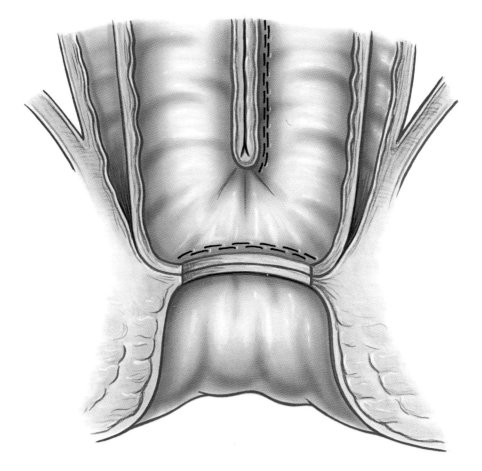

Figure V-30,G: The anastomosis can easily be inspected transanally; it is composed of a full-thickness wall on the ileal side and of mucosa and tough submucosa on the anal side.

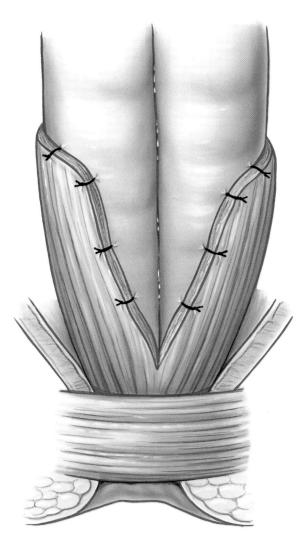

Figure V-30,H: The ileal pouch is anchored to the rectal muscularis with interrupted sutures to hold it in place and protect the anastomosis from disruption, as the patient is allowed to straighten out progressively after the operation.

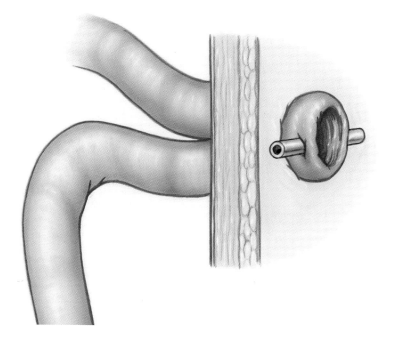

Figure V-30,I: A proximal diverting ileostomy is established to allow this atypical anastomosis, which is often under some tension, to heal.

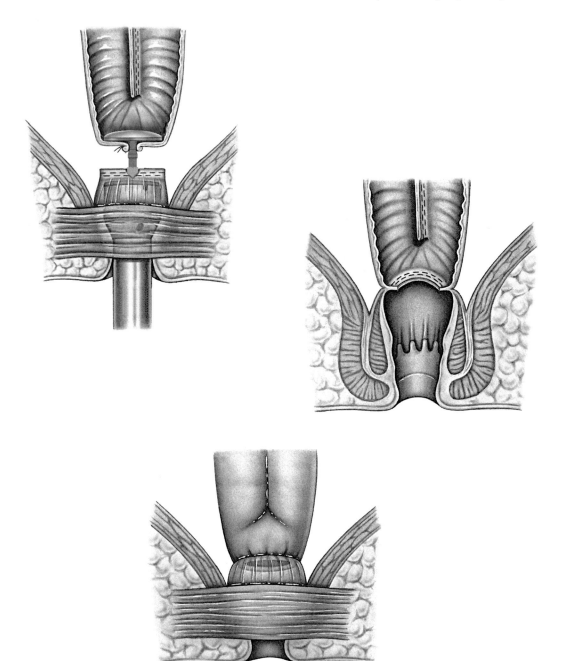

Figure V-30,J,K,L: An alternative solution to satisfy the requirement for the pouch to extend deep into the pelvis and avoid tension on what is a somewhat tenuous anastomosis, consists in preserving the anal canal and anastomosing the pouch to a full-thickness anorectal junction, using the double stapling technique. This solution does imply the existence of a 2 to 3 cm mucosal collar above the dentate line that has a potential for malignant transformation in patients with ulcerative colitis and familial polyposis, albeit at a level where regular inspection, biopsies, and remedial action as, and if, necessary are easily accomplished and may represent a small price to pay for a safer pouch to anus anastomosis.

LOW ANTERIOR ABDOMINAL DISSECTION COMBINED WITH RESECTION AND COLOANAL ANASTOMOSIS THROUGH AN ANTERIOR PERINEAL APPROACH

This procedure combines the advantages of a clear and wide access to state-of-the-art dissection of the rectum through an abdominal approach and of a similarly secure, controlled exposure for proctectomy and anastomosis of the sigmoid colon to the anus, through an anterior perineal approach. It does require an abdominoperineal sequence, with the perineal stage depending on placing the patient into an exaggerated lithotomy position: thighs abducted and flexed onto the pelvis, stretching the perineum transversely and tilting it anterosuperiorly into a horizontal plane, as viewed from above. By dropping the patient's shoulders and trunk below the horizontal line of a normal dorsal decubitus, into a steep inclination, the anterosuperior tilt is accomplished. However, this does require careful precautions against leg vein stasis and thrombosis and padded and safe shoulder guards and is not suited for patients with cardiac and pulmonary deficiencies (Welter, Psalmon, 1987). While the anterior perineal approach is used to facilitate a very low anterior coloanal anastomosis, it does not preclude the conversion to a true abdominoperineal amputation should intraoperative findings demonstrate that reconstruction of continuity would be to the detriment of the patient's long-term survival. Therefore, patients should be prepared preoperatively and have given consent for both outcomes.

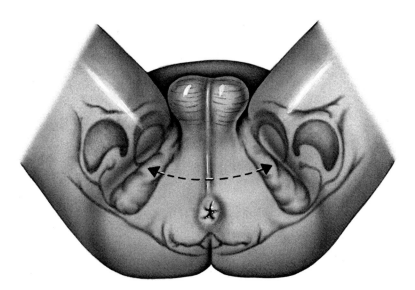

Figure V-31,A: The transverse perineal skin incision is placed 2 cm in front of the anus. The anobulbar raphe is divided, exposing the anterior aspect of the rectum, just above the external sphincter and deep to the levator ani muscles on both sides. It is then easy to pass an index finger and thumb by blunt dissection around the rectum, through the surrounding areolar tissue, by staying more distal on the dorsal side, in order to free the circumference of the rectum in the same plane. A rubber drain is passed around the rectum to facilitate exposure. In the elderly patient, the weakened levator ani muscles can usually be retracted, whereas in younger patients, strong and fleshy levator muscles may require partial section on one side and later repair, to provide enough exposure. At this point in the procedure, the decision is made as to the feasibility of anterior resection and coloanal anastomosis, versus abdominoperineal amputation, by measuring the safe margin beyond the tumor. It should at least be 3 cm between the lower pole of the tumor and the upper aspect of the striated sphincter, to favor anastomosis over amputation.

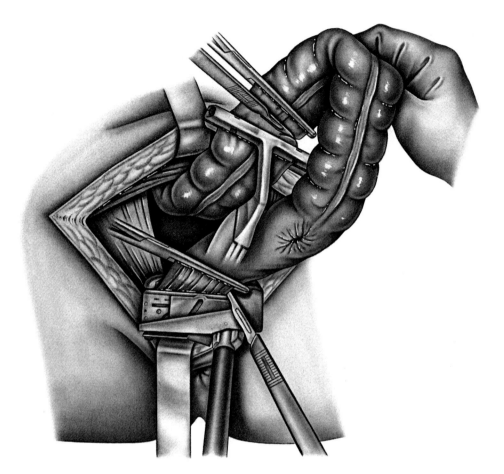

Figure V-31,B: With gentle traction on the rubber drain around the rectum, the abdominal dissection is completed from below. The tumor-bearing rectum is pulled down, through the perineotomy, until healthy sigmoid colon with intact blood supply reaches the upper aperture of the perineum. At this stage of the procedure, the rectum is closed with the Roticulator instrument above the striated sphincter (the range can be up to 2 cm above this level, depending on the safe margin to the tumor). The rectum is transected between the distal linear stapler, using its upper edge as a guide, and a proximal Kocher clamp, occluding the rectum on the specimen side. The Roticulator instrument is removed. On the sigmoid side of the specimen, the Purstring instrument is placed onto healthy viable bowel and is activated to place a purse-string suture. The sigmoid is then transected between the Purstring instrument and a clamp, occluding the specimen, this time by using the lower edge of the Purstring instrument as a guide.

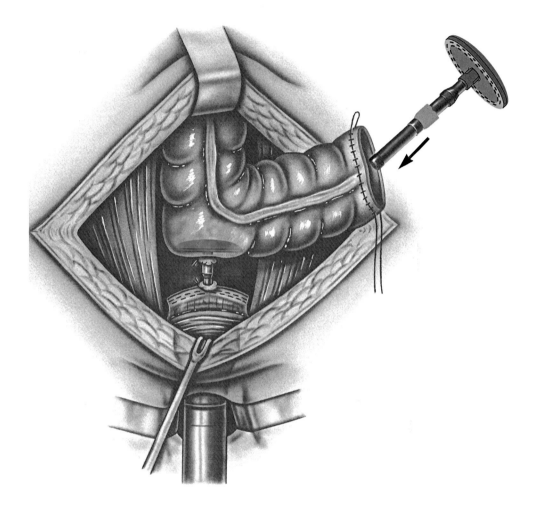

Figure V-31,C: The specimen is removed and examined one more time for adequate tumor margins. If they are unsatisfactory, the abdominoperineal amputation is completed by resecting the anus with the usual perineal collar of skin, subcutaneous fat, and muscles. If satisfactory, the anastomosis is now started by placing the EEA instrument, with cartridge only, into the anal canal and advancing the center rod through or near the center of the linear staple line. The anvil is passed into the open proximal sigmoid colon and secured with its shaft leading and brought out through a stab wound in the antimesosigmoid wall of the normally curved sigmoid that approaches the anal closure with the least tension.

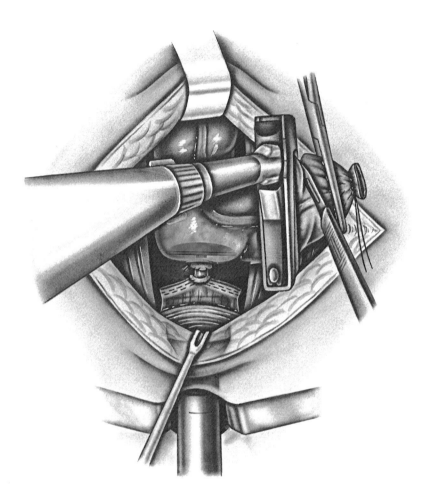

Figure V-31,D: The EEA anvil and cartridge are closed against each other and the instrument is activated, producing a side-to-end sigmoid-anal anastomosis. The excess sigmoid colon beyond the lateral border of this anastomosis is closed with the linear stapler and excised using the Roticulator or TA stapler used as a guide for the scalpel.

Figure V-31,E: The final result is a side-to-end sigmoid-anal anastomosis, following the pattern of the double stapling technique. By using the antimesocolic side of the normally curved sigmoid colon, the diameter of the bowel adds some 3 to 4 cm to the length of the afferent limb and relieves some of the tension that often complicates such extremely low anastomoses. Since we have used this side-to-end anastomosis rather than a true end-to-end sigmoid-anal anastomosis, postoperative strictures have become an experience of earlier times.

ILEAL POUCH TO ANAL CANAL ANASTOMOSIS THROUGH AN ANTERIOR PERINEAL APPROACH

Nowhere is the utility of the abdominal and anterior perineal sequence more apparent than in the combined operation of total colectomy and creation of a substitute ileal pouch. Each exposure fulfills a specific purpose: clear and accessible anatomy to ensure a safe colon dissection and mobilization, as well as state-of-the-art construction of the substitute pouch through the abdomen, and secure and reliable pouch to anus anastomosis under direct vision through the anterior perineum.

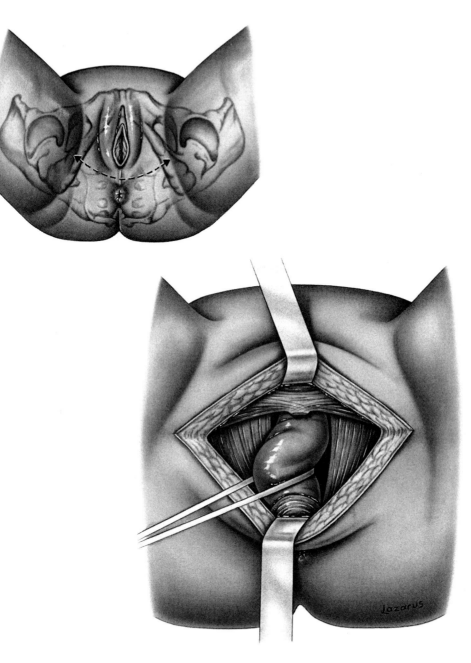

Figure V-32,A,B: Following transabdominal dissection and liberation of the colon and rectum, as well as construction of an ileal J-pouch, the perineal stage of the operation— positioning of the patient and exposure of the rectum— is performed as previously described for Figure V-31,A,B, with the exception that in this illustration, the anovulvar raphe would be transected after the anterior transverse perineal skin incision has been accomplished.

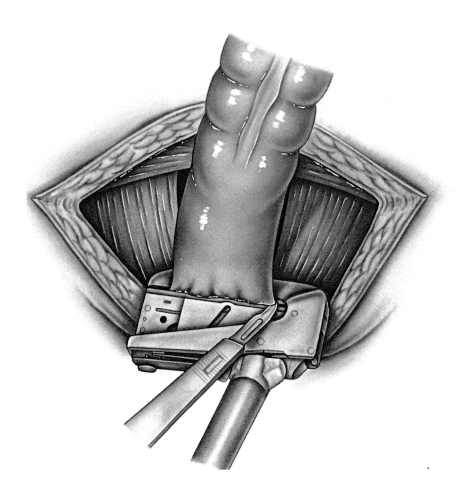

Figure V-32,C: The entire specimen, consisting of colon and rectum, is pulled through the anterior perineotomy and the rectum to anal canal junction is closed and stapled transversely with the linear Roticulator or TA instrument, while holding the specimen vertically to facilitate this maneuver. The bowel is transected, using the upper edge of the instrument as a convenient guide.

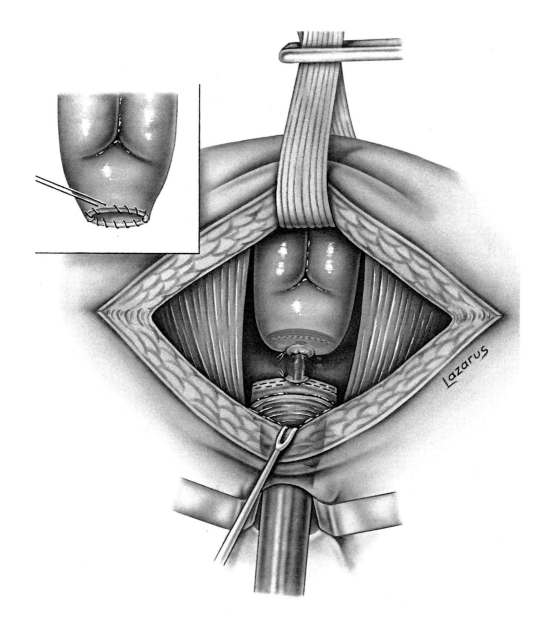

Figure V-32,D: The previously constructed J-pouch is brought down and a purse-string suture is placed around the common opening left by the use of the GIA side-to-side anastomosing instrument.

Figure V-32,E: A wide corrugated drain, passed from the abdomen around the genitourinary organs and the pubis and clamped above the abdominal wall, holds the genitourinary organs out of the way and provides a clear view into the pelvis. As previously shown (Figure V-30,F), the anvil is placed into the J-pouch and the purse-string is tied around the anvil shaft. The EEA instrument, with the cartridge only, is positioned into the anus, against the linear closure and the center rod is advanced through the linear closure. The anvil shaft and center rod are joined, and the anvil and cartridge are closed against each other.

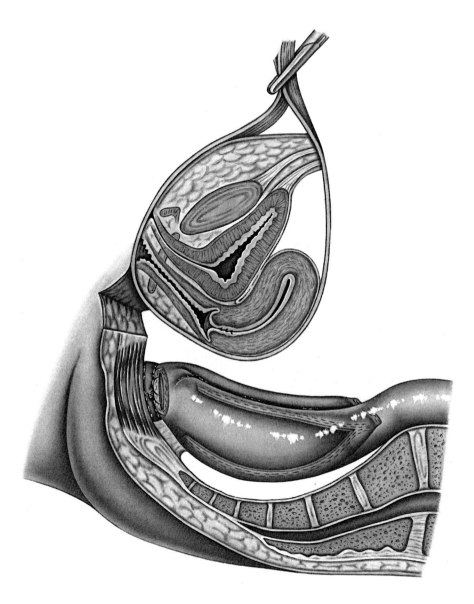

Figure V-32,F: The EEA instrument is activated and a pouch to anal canal anastomosis created. The wide corrugated drain does keep the anterior pelvic organs— in this case, the bladder and the uterus— out of the way.

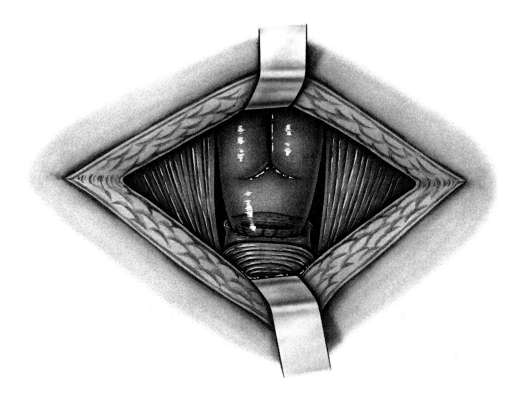

Figure V-32,G: The final result is inspected to examine the state of the anastomosis, just above the striated sphincter muscle. The easy access through the perineotomy would facilitate any necessary remedial action.

ILEAL POUCH TO ANUS ANASTOMOSIS THROUGH AN ANTERIOR PERINEAL APPROACH

The anterior perineal approach, with its easy and direct access to the lower rectum, facilitates to a great measure the mucosectomy required to eliminate all potential for malignant transformation in patients who are undergoing near-total coloproctectomy for ulcerative colitis and familial polyposis and who, along with the surgeon, prefer to obviate the constraint for more frequent endoscopy and biopsy of the small collar of mucosa left above the dentate line, as shown in the preceding operation.

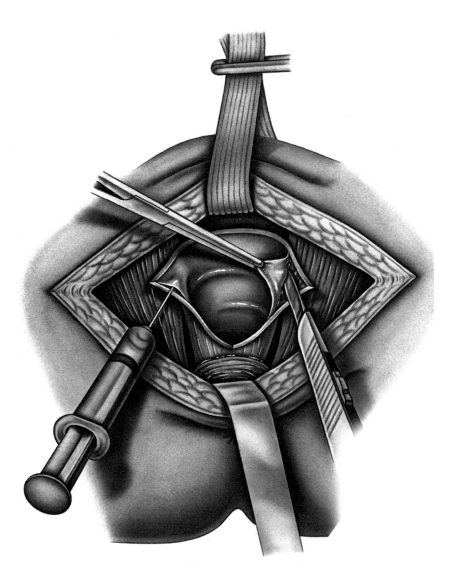

Figure V-33,A: As shown repeatedly before, the rectum has been exposed through the now familiar anterior perineotomy and transected, leaving a rectal stump of some 5 to 6 cm above the striated sphincter muscle. The anterior wall of the rectal stump is incised vertically, down to the sphincter. The mucosectomy is started at both corners of the book-like opening of the rectal cylinder, by injecting saline to elevate mucosa and submucosa, to facilitate the separation of both from the muscle layers by sharp dissection.

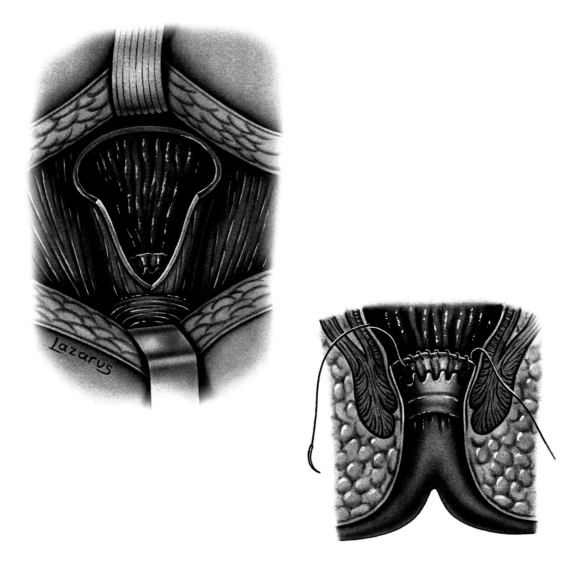

Figure V-33,B: The mucosectomy is extended to a level, some 0.5 cm above the dentate line.

Figure V-33,C: The purse-string suture is placed into the circular rim of mucosa and tough submucosa left at this level, with a manual over-and-over whipstitch.

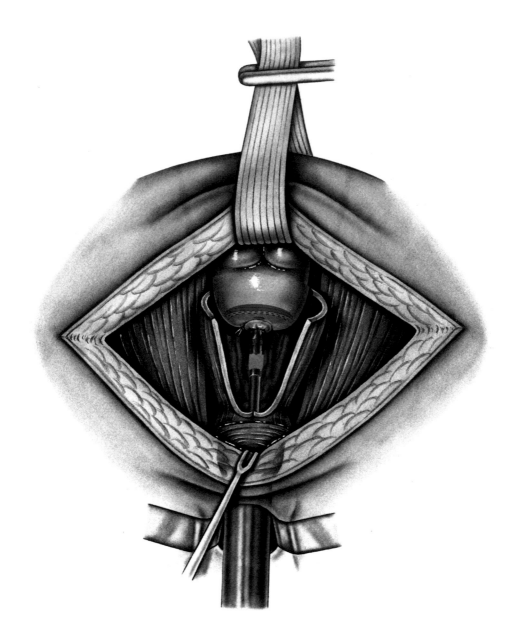

Figure V-33,D: The broad corrugated drain around the genitourinary organs and the pubis keeps the approach clear and makes for an easier transfer of the previously constructed pouch from the abdomen into the pelvis. As before, the anvil is placed into the pouch opening and that purse-string is tied around the anvil shaft. Next, the EEA instrument, with cartridge, is advanced into the anus and the mucosa-submucosa purse-string is tightened around the center rod. Both anvil shaft and center rod are mated.

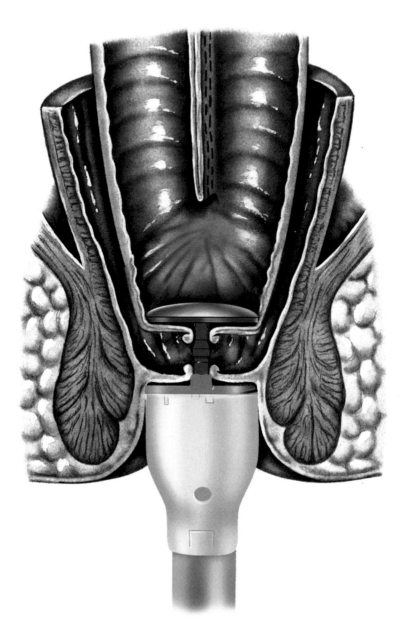

Figure V-33,E: The anvil is closed against the cartridge, taking great care to help along the sliding of the pouch into the gaping muscular rectal stump.

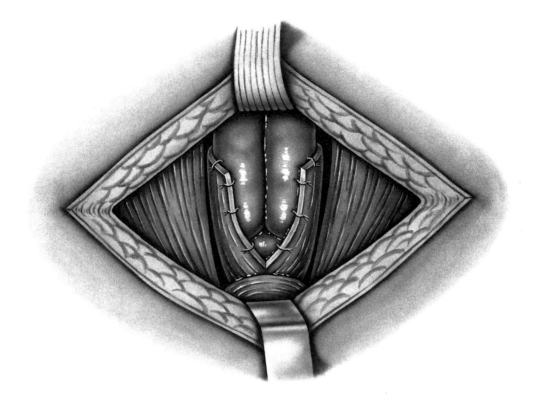

Figure V-33,F: After the pouch to anus anastomosis has been accomplished, the pouch is anchored into the rectal stump with sutures from the pouch to the borders of the stump, along the now V-shaped vertical incision and around its proximal circular contour.

COLON POUCH TO ANUS ANASTOMOSIS THROUGH AN ANTERIOR PERINEAL APPROACH

Conventional teaching in the case of a "routine" anterior resection was that the pulled down sigmoid and/or descending colon would in time substitute for rectal function and return the patient's bowel habits to normal or near normal. While this continued to be true with the use of the end-to-end, circular anastomosing instrument in anterior resections done at commonly average levels, in the very low anastomoses the rehabilitation to near normal bowel habits became uncommonly frustrating and drawn out. The usual suspects, excessive retraction of the anal sphincters by the assistants or now-prolonged distension of the anus by the bulky EEA instrument, were considered. But it soon became obvious that of the many factors that regulate anorectal continence and that are all "visited" by a total mesorectal and rectal excision—nerves, vessels, pelvic floor and sphincter muscles—the storage capacity of the rectum was affected most by the complete elimination of this organ, made possible by this new, relatively facile and secure, very low anastomosis. Therefore, the suggestion was made (Fazio, 1984) to create a colon pouch in patients with low anterior resection and sphincter preservation, along the model of the ileal pouch. Serendipitously, we had already noted better healing of side-to-end coloanal anstomoses with the technique shown in Figure V-31, A to E. This was attributed to the reduction of tension, and hence better blood supply, by the use of the lateral aspect of the colon for anastomosis, relaxing the pull exercised by the mesocolon. A concomitant improvement in bowel habits was also attributed to better healing without stenosis, when in fact, at the time we created a mini-pouch by adopting the golf club or hockey stick configuration of the side-to-end anastomosis.

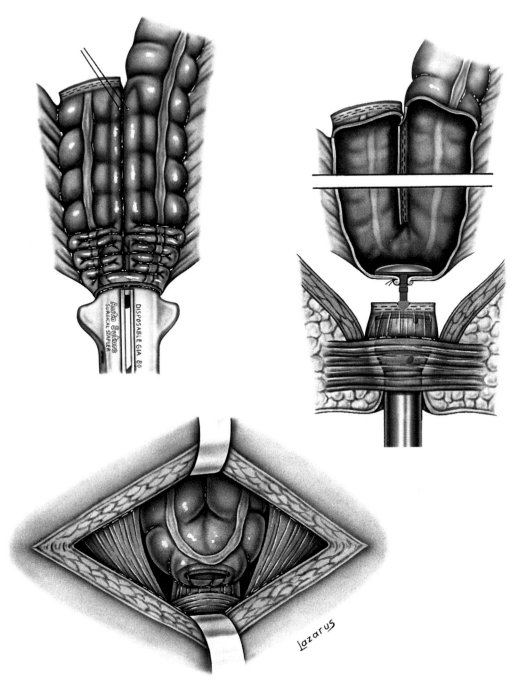

Figure V-34,A: The colon pouch is created in the now familiar fashion, with two applications of the GIA instrument (Figure V-30, A to C).

Figure V-34,B: The anastomosis of the pouch to the transversely stapled anus is performed using the double stapling technique (Figure V-21, A to F).

Figure V-34,C: A safe neo-rectal reservoir is created, while the sphincter muscles are preserved and the pelvic floor is reconstructed.

DUHAMEL OPERATION FOR CONGENITAL MEGACOLON - HIRSCHSPRUNG'S DISEASE

When Mark Ravitch returned in the fall of 1958 from his trip to Russia with the UKB instrument for bronchial stapling, he had seen at the Institute in Moscow the Russian precursor of the GIA instrument, the NZhKA, adult and pediatric models. His wide-ranging interests in all things surgical, including pediatric surgery, led him to predict that this instrument would greatly simplify the transection and side-to-side suturing of the colorectal spur of the Duhamel procedure if the instrument's handling could be simplified and the resulting tissue transection and simultaneous side-to-side staple suturing be made more reliable. By then we had had some experience with manually performed Duhamel procedures and had always been dismayed, as were the parents, nurses, and probably our young patients, with the need to let a heavy clamp protrude through the anus postoperatively, to crush the colorectal spur, of what appeared otherwise as a very ingenious and elegant solution to a very difficult congenital anomaly.

Together with Jim Talbert, Ravitch did some experimental work to prove his point, first with the Russian instrument and then with the GIA stapler as it became available. In 1968, we published our initial clinical experience with the technique as will be shown here, except that the availability since then of the EEA instrument allows for a completely "mechanized" procedure. The limitation being the size of the instruments that reduce the utility of this technique to children, 9 to 12 months old and up, even if the smallest EEA instrument and today's ENDO GIA stapler are used.

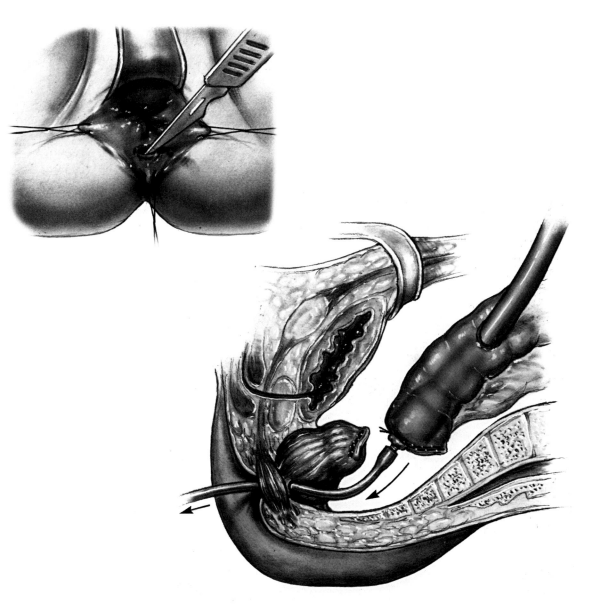

Figure V-35,A: After the rectum has been stapled closed and divided below the peritoneal reflexion and the aganglionic segment of colon removed, a transverse stab wound is placed into the posterior wall of the rectum, just above the sphincter and reaching into the retrorectal space, previously opened from above.

Figure V-35,B: The EEA 21 instrument, without anvil, is placed into the proximal, healthy colon and the center rod of the instrument is advanced through its stapled closure, reinforced by a purse-string suture around the center rod. The funnel of a red rubber catheter is attached to the protruding rod and the catheter is used to guide the proximal colon, containing the EEA cartridge and center rod, behind the rectum toward the posterior rectal stab wound and the center rod through this stab wound.

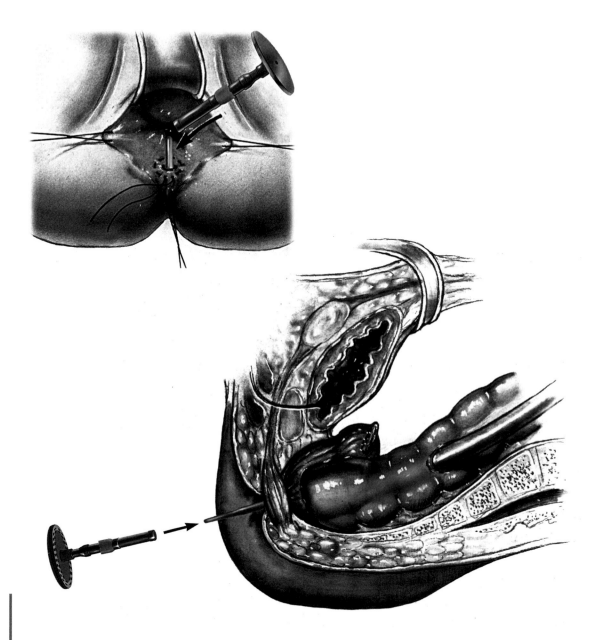

Figure V-35,C: The catheter is removed from the center rod, a purse-string suture is placed around it in the rectal wall, and the EEA anvil is attached to the center rod.

Figure V-35,D: The future circular, end-to-side, low colorectal anastomosis with the EEA instrument is shown in a lateral view.

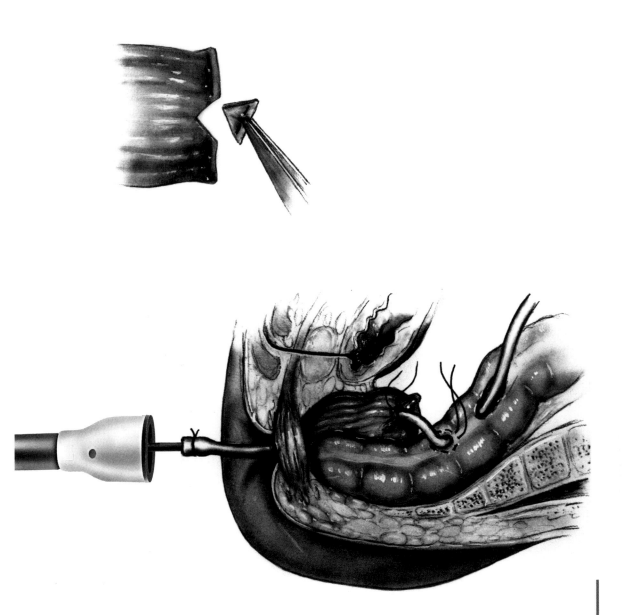

Figure V-35,E: Next, a small cuneiform excision of the center of the rectal staple closure is done.

Figure V-35,F: The red rubber catheter is again used as a guide, in the opposite direction, inside the rectum, out through the above cuneiform excision and through a kissing stab wound into the proximal colon. It is retrieved through the previous proximal colotomy. The center rod of a new EEA 21 instrument is attached to the funnel of the red rubber catheter outside the anus. The passages of the red rubber catheter, out the rectum and into the colon are narrowed with purse-string sutures that are left untied.

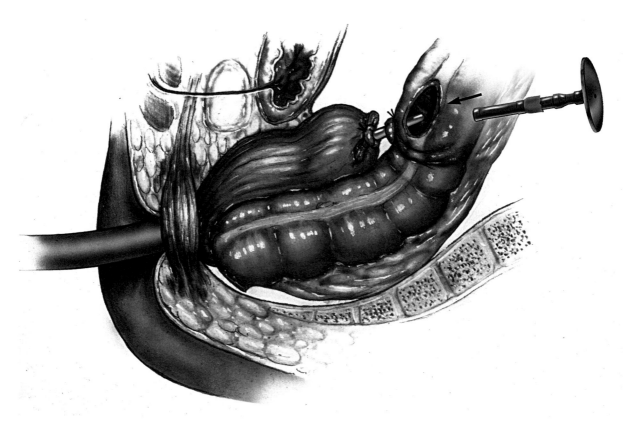

Figure V-35,G: As the red rubber catheter is moved superiorly, the cartridge of the EEA instrument follows and is positioned below the rectal closure. The center rod traverses the opening in the rectum and the stab wound in the proximal colon. The purse-string sutures are tightened at this point and the anvil shaft is attached to the center rod through the previous proximal colotomy.

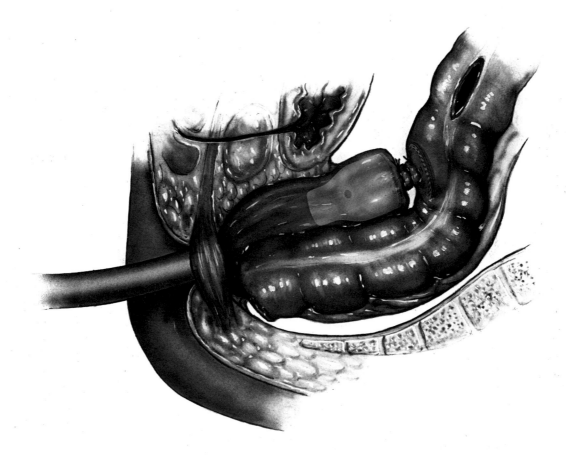

Figure V-35,H: Anvil and cartridge are closed against each other and the proximal side-to-end anastomosis is accomplished. The circular "donuts" of excised tissue remaining within the instrument are examined after each EEA instrument application.

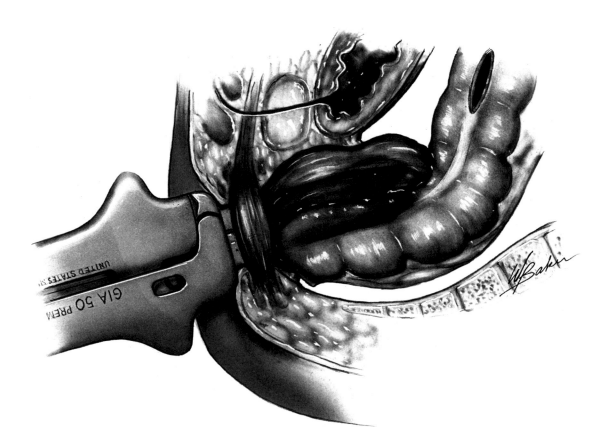

Figure V-35,I: The colorectal spur is now delineated by both the superior and inferior EEA colorectal anastomoses. The GIA stapler or, if preferable because of patient size, the ENDO GIA instrument is now used to connect the two circular anastomoses. Depending on the length of the spur, two applications may be required. This step demonstrates the safe transection of circular staple lines by linear ones.

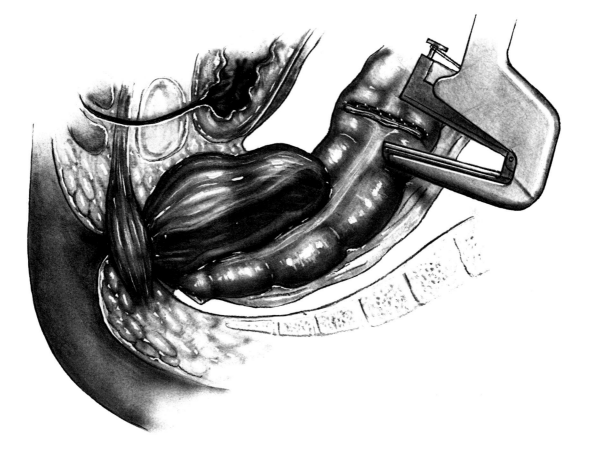

Figure V-35,J: All anastomoses are observed for hemostasis, and the proximal colotomy is closed transversely with a linear stapler.

REPAIR OF RECTOVAGINAL FISTULA

A rectovaginal fistula, usually the result of radiation treatment for carcinoma of the uterus, at other times the consequence of an operative or postpartum injury, reduces the life of a woman—often young, never beyond active middle age—to an existence of anguish, social isolation, marital separation, deep desperation, and near-suicidal depression. The tragedy of its occurrence lies in the fact that the condition for which the ultimately injurious treatment was undertaken, as for instance cancer of the uterus treated by hysterectomy and radiotherapy, has been cured. But the patient is left with a histologically benign condition that affects her life almost worse than a cancer and that is the erratic complication of a well-planned and competently executed treatment. These patients suffer from constant fecal discharge through the vagina, requiring the wearing and frequent changing of hygienic pads. Their perineum and buttocks are red and irritated with inflammation, often the site of micro-abscesses. It is tempting to remedy this miserable state of affairs by excision of the fistula and surrounding scar tissue, primary closure of the vagina, anterior coloanal anastomosis, and interposition between vagina and low anastomosis of the always cited, but never present (anymore at this stage) elongated greater omentum. However, we have been concerned with the poor state of the vasculature deep in the pelvis that led to the wretched condition in the first place. The remedial operation proposed by Bricker and Johnston, consisting, in essence, of bringing down the healthy sigmoid colon as a graft, with vascular pedicle, to close the anterior rectal wall and act as a buffer between rectum and vagina, appeared to us as the better solution, causing minimal local tissue trauma and bringing new blood supply to the area. Since this is a multi-stage, technically involved procedure, we have adapted the use of staplers to its construction in order to shorten operative time and provide the security of mechanical sutures.

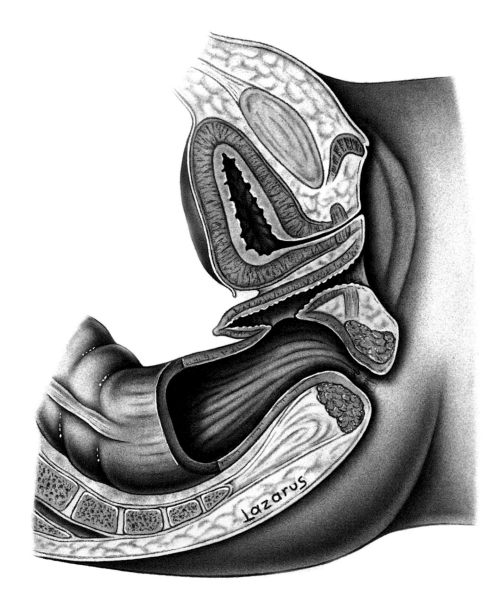

First Stage: Healing of perineal skin inflammation and infection.

Figure V-36,A: The diagnosis is usually very obvious and does not require many additional clinical or laboratory examinations, except for the need to perform endoscopies of rectum and vagina and to obtain four quadrant biopsies to rule out malignancy, in which case, a permanent colostomy and abdominoperineal resection of rectum and vagina would be indicated.

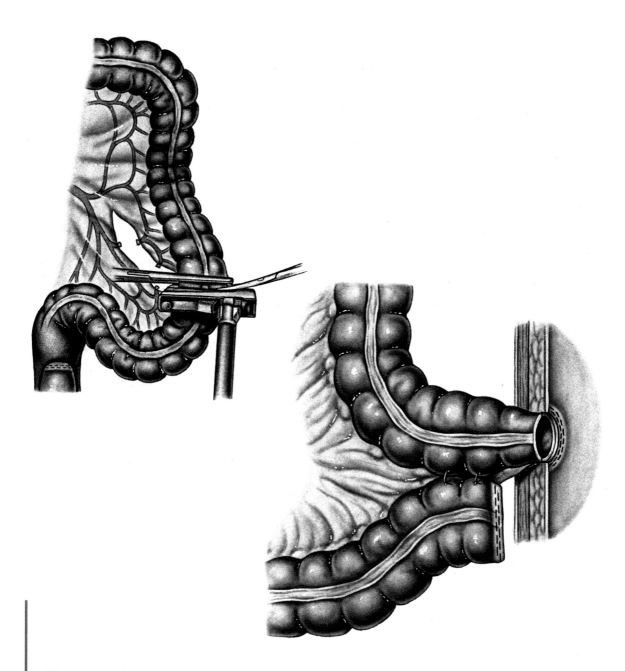

Figure V-36,B,C: After the benign state of a given fistula has been confirmed, a diverting colostomy is created from the lower descending colon and the sigmoid graft is prepared by closing the colosigmoid junction with a linear stapler, attaching this closure to the colostomy inside the peritoneal cavity and carefully separating the blood supply to the sigmoid from the one supporting the descending colon. The fully diverting colostomy serves to put the perineal skin infection at rest and to prepare for a second, abdominoperineal stage, free of any complicating local factors.

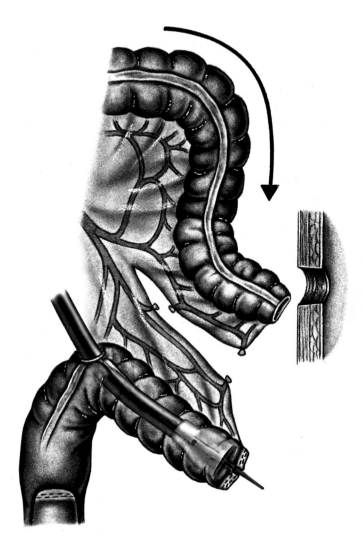

Second Stage: Abdominoperineal repair of fistula with sigmoid graft carried by vascular pedicle. After the perineal skin has healed and is clear of all infection, the second stage of the procedure is undertaken.

Figure V-36,D: Through an abdominal approach, the descending colostomy is dissected from the abdominal wall. The sigmoid and descending colon, together with their previously established individual vascular pedicles, are separated. The future sigmoid graft is assessed for mobility, reach of pedicle and blood supply, and optimal introduction site of the EEA instrument, without anvil. This site is usually at the apex of the sigmoid-rectal junction.

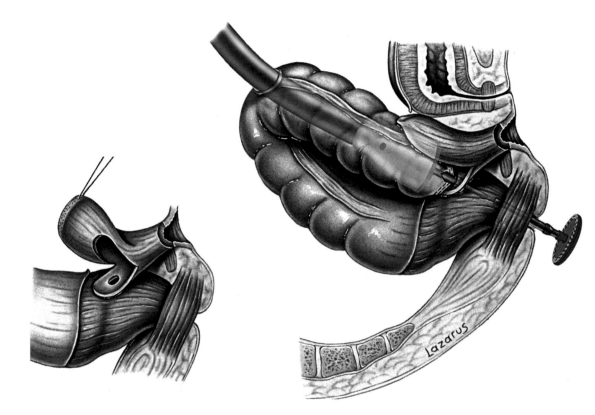

Figure V-36,E: The vaginal cuff is now lifted off the anterior wall of the rectum by sharp dissection, and the fistulous connection is divided by leaving a collar of tough fibrous tissue around the fistula on the rectum. This collar will act as a purse-string suture that would be difficult to place at this level by hand or by machine.

Figure V-36,F: The EEA instrument, without anvil, is placed into the previously prepared sigmoid graft, at or near the rectosigmoid junction, ideally at the half-way apex of the rectosigmoid segment. The EEA cartridge is moved toward the proximal sigmoid closure (Figure V-36,D) and its center rod is advanced through the closure and secured with a purse-string suture. With the shaft and handle of the EEA instrument used as a carrying stick, the sigmoid colon is transferred deep into the pelvis, its proximal end bolstered by the EEA cartridge, leading the way. The center rod of the instrument is maneuvered through the rectal aperture of the fistula, from outside to inside the rectal lumen and is extended to the vicinity of or outside the anus. The anvil is then attached to the center rod. The anvil and cartridge are closed against each other and the end-to-side proximal sigmoid to rectum anastomosis is accomplished.

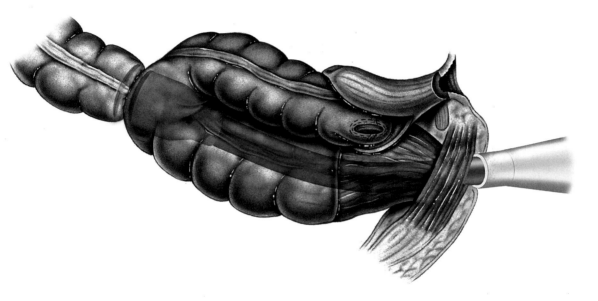

Figure V-36,G: Bowel continuity is re-established by anastomosing the distal descending colon to the apex of the rectosigmoid segment, in an end-to-side mode, with the EEA instrument introduced through the rectum.

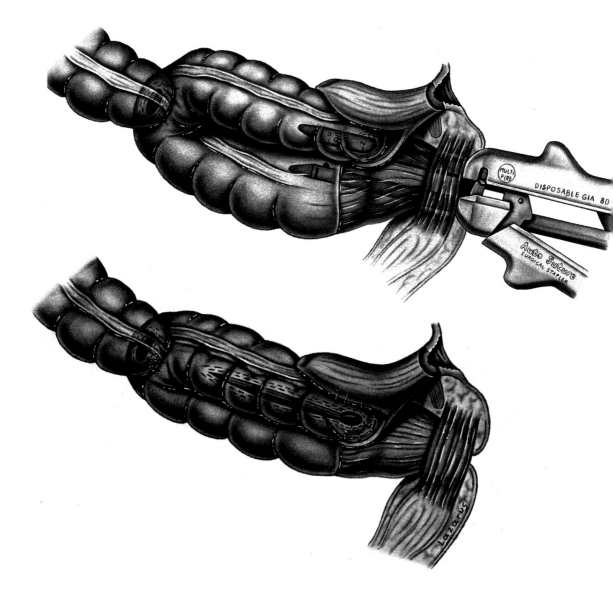

Figure V-36,H: As in the Duhamel operation, the surgeon is now faced with a colorectal septum or spur, albeit this time on the anterior surface of the rectum. However, the solution of dividing and stapling side-to-side both walls of the spur with a long GIA instrument remains the same; the exception being that one arm of the instrument is placed into the posterior rectum and the other arm through the anus and previous fistula, into the anterior sigmoid colon, and that the division of the spur does not have to be complete (all the way into the proximal colon).

Figure V-36,I: The division of the spur should involve at least 70% to 80% of its length. As in the Duhamel operation, this shows the safe transection of an annular staple line by a linear one.

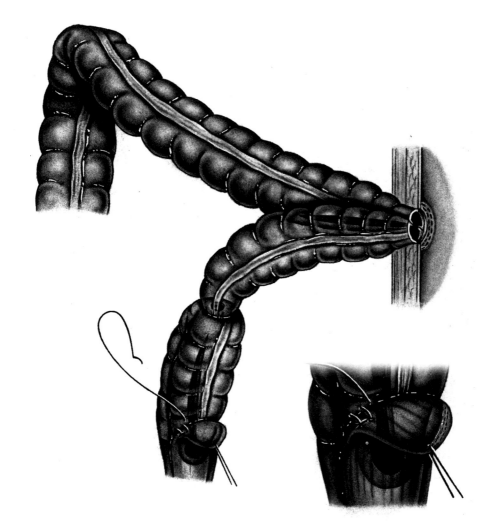

Figure V-36,J,K: To protect the major anatomical rearrangement that this operation represents, a vented, proximal colocolostomy is performed (Figure V-17, A to H).

The vaginal cuff is loosely sutured to the anterior sigmoid wall, first around the refreshed fistulous aperture and then around the contour of the cuff. In this way, a portion of the posterior vaginal wall is represented by the serosa of the sigmoid colon, which in time assumes the same gross aspect as the vaginal mucosa. While initially we were very concerned about this exposure of bowel wall, we have so far, in an admittedly and fortunately small number of patients, observed no breakdown of the sigmoid colon, even though some of the younger patients have returned to cautious sexual intercourse.

Third Stage: After all anastomoses and incisions have healed and normal bowel habits have returned via rectum and anus, the vented colocolostomy is closed as shown in Figure V-17,J,K,L).

MECHANICAL SUTURES IN SMALL AND LARGE BOWEL OPERATIONS BY LAPAROSCOPY

Since the first appendectomy by laparoscopy in 1980 by Kurt Semm in Kiel, Germany, and the first laparoscopic cholecystectomy by Philippe Mouret in 1987 in Lyon, France, the movement to extend laparoscopic and thoracoscopic procedures beyond diagnostic explorations and single action therapeutic procedures in the chest and the female pelvis, performed by gynecologists and pneumologists mostly up to that point, gained more and more momentum on a global scale. Beyond the simplistic snob appeal and the one-upmanship of a new and exciting technical approach to excision only, or excision and reconstruction, or repair only, it soon became obvious that the traditional incisions and intracavitary exposures represent a large component of the overall trauma inflicted by a conventional surgical operation. And the technical details of intracavitary, endoscopic, full-fledged operative procedures, without a conventional incision of some sort or other, became the foundation and building blocks for the principle of Minimally Invasive Surgery; minimally invasive not only of the patient's anatomy, but also of his or her immune response to infection and neoplasia, minimal interference with a variety of postoperative organ functions, and minimal abuse of personal and public financial resources. Laparoscopy and thoracoscopy are techniques; Minimally Invasive Surgery is a concept, made possible in part by the two techniques, but also by gentle tissue handling and careful hemostasis (Halsted), avoidance of intraoperative contamination and infection, expert maintenance intraoperatively and sensible resuscitation postoperatively of all vital functions, compassion for pain and fear, and so many incidental parameters, which surgeons have striven to satisfy since ancient times.

The distance covered in 24 years is best illustrated by the fact that in 1980, Karl Semm was almost reprimanded by his colleagues of the Kiel University faculty for having used a proven gynecological tool in an unusual application above the pelvic rim to an organ that was considered to be part of the operative repertoire of a different specialty. In fact, the first practice of cavitary endoscopy in gynecological "pelviscopy" had been just as empirical and successful as in Semm's "celioscopy." The quest in the healing arts to see normal and abnormal states and to believe in a diagnosis based on visual findings, which leads to a reasonable therapeutic approach, goes back to the speculum first found in ancient Pompeii and is highlighted by such milestones as the rigid, and later flexible, cystoscope, bronchoscope, esophagogastroscope and sigmoidocolonoscope, as well as the "Roentgen x-rays" and all the powerful imaging means developed since 1895.

Ironically, in today's world, we have to be concerned about the opposite of Semm's reception by his peers, namely the witless and freewheeling use of endoscopic techniques first and foremost, when in fact the choice between laparotomy and laparoscopy for instance, as a primary approach, should always be guided by the Hippocratic aphorism of "Primum Non Nocere." The case of acute appendicitis, clinically strongly suspected and sonographically confirmed in a male child, can serve once more as an example; after initial enthusiasm for laparoscopic appendectomy, it soon became obvious that in fact the endoscopic approach did not present any tangible advantages over appendectomy through a buttonhole McBurney incision. However, this "ex aequo" position was achieved at a greater cost for laparoscopy and the always present risk of a trocar injury. It should be added that the laparoscopic experience in this case, however, did teach the valuable lesson of early discharge after operation; a lesson that was then applied to conventional appendectomy without any ill-effect, resulting in savings in both examples. This goes to show only that traditional operative approaches, when indicated, should not convey the aura of hopeless obsolescence and that a constant comparison between all the factors involved in an open versus a closed approach should allow a reasonable decision-making process, as demonstrated for acute appendicitis in a male child.

Very early in the development of modern thoracoscopy and laparoscopy, it became quite clear that there was a symbiotic relationship between cavitary endoscopy and mechanical sutures. Many of the operative techniques previously shown (Figure V-1 to Figure V-35) can be transferred, in toto or in part, to the reconstructive phase after laparoscopic bowel resections. However, in contrast to excision only of an organ or structure by laparoscopy, the need for reconstruction after bowel resection presented a new set of problems, namely the requirement of access larger than was available through the average, or even supersized, trocar site. This challenge was met by a compromise; a change from the pure, laparoscopically guided (best example being cholecystectomy), to the laparoscopically assisted operation, allowing a small incision through which to remove the specimen and perform anastomosis. This challenge also reintroduced the functional end-to-end anastomosis and its varieties to their right places for intraperitoneal anastomoses, reserving the circular end-to-end anastomosing instrument for use through the anus, a natural body orifice, thus avoiding the enterotomy that is required if this instrument is used above the pelvis.

The minimally invasive approach required a totally new mode of access techniques to the abdomen, assured by trocars, as well as new means of exposure, provided by pneumoperitoneum. While the placement sites of the trocars for access were initially very rigidly defined for each operation, time, patient habitus, pathological findings, surgeon's experience and sound imagination, and available instrumentation have created some flexibility in the choice of the various access sites. Added to this choice in access sites are the innumerable trocars and dissection and hemostasis instruments, available at present and still on the drawing boards. We, therefore, refer the reader to the many specialized texts that provide this information and will only show here the participation of mechanical sutures in the context of laparoscopic operations.

RIGHT HEMICOLECTOMY AND ILEO-TRANSVERSE COLOSTOMY

The anatomical side-to-side and functional end-to-end anastomosis lends itself well to an intraabdominal anastomosis with preservation of the pneumoperitoneum. However, the same reasoning that guides the decision making for appendectomy may prevail for a right hemicolectomy performed for a benign condition or a small malignancy in the cecum or ascending colon of an elderly patient with a thin, relaxed abdominal wall. The conventional right lateral, transverse, or oblique incision, may not present a greater affront to the patient's well being than the combined extent of all trocar sites and abdominal incision for anastomosis and evacuation of the specimen, as well as the increase in operating time and related expense. Paradoxically, the opposite finding, a bulky cecal tumor, may very wisely lead to conversion into an open operation, early in an attempt at laparoscopy, because of the technical limitations imposed by the confines of the pneumoperitoneum.

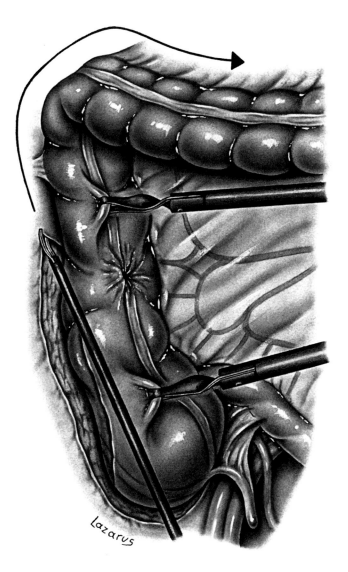

Figure VL-1,A: After satisfactory laparoscopic exposure has been obtained and the diagnosis has been confirmed visually, the parietal peritoneum is incised and the cecum and ascending colon to be resected are freed by sharp dissection and progressive elevation toward the midline. Additionally, the hepatocolic ligament at the hepatic flexure can be transected with an application of the ENDO GIA instrument.

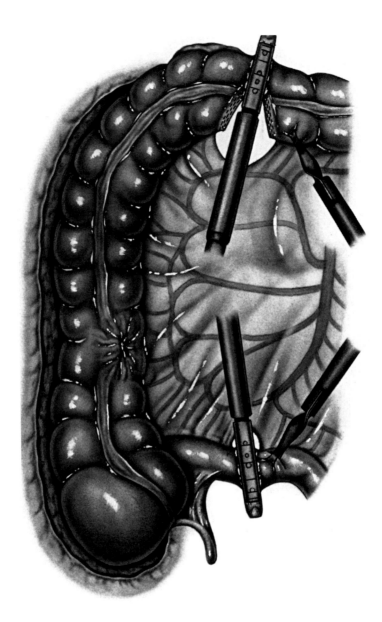

Figure VL-1,B: Windows are created in the mesentery and mesocolon at the projected points of transection of the distal ileum and transverse colon. The contour of the future specimen is outlined between applications of the ENDO GIA instrument, which simultaneously places three staggered rows of staples on each side of the transection, to close both the specimen and the remaining bowel ends. On each application, care is taken to prevent inclusion of extraneous tissue or structures within the confines of the instrument prior to closure.

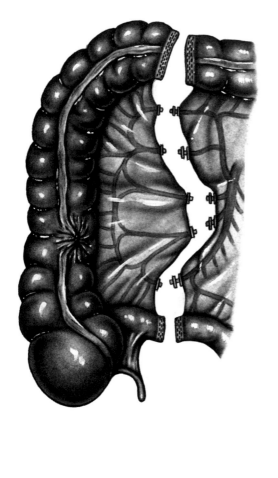

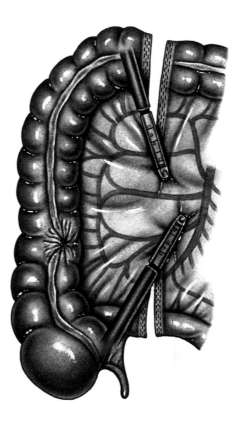

Figure VL-1,C: The ileocolic and right colic vessels, at their respective origins, are doubly ligated on the patient side and singularly ligated on the specimen side to complete the resection.

Figure VL-1,D: Alternatively, the terminal ileal mesentery and right mesocolon can be separated and transected with serial applications of the ENDO GIA instrument, placed close to the origins of the various vessels comprised in the specimen. Appropriate staple size should be chosen for the thin mesenteric tissue compared to those required for the thicker bowel tissue. It should be noted that steps C and D are facilitated by the transillumination provided by the endoscope, especially in thin patients. Furthermore, this transillumination will secure early vessel controls in the "no-touch" technique for malignant disease, before mobilization of the bowel.

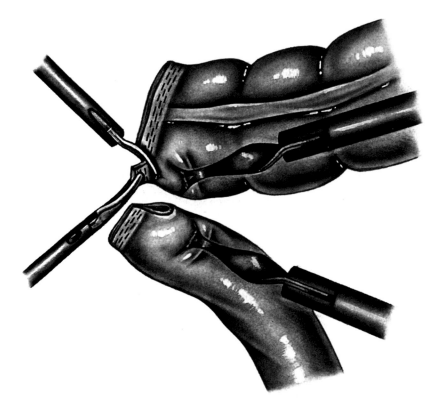

Figure VL-1,E: The specimen is temporarily stored in the right lower quadrant, and the antimesenteric corners of the stapled closures of ileum and proximal transverse colon are excised on the patient's side.

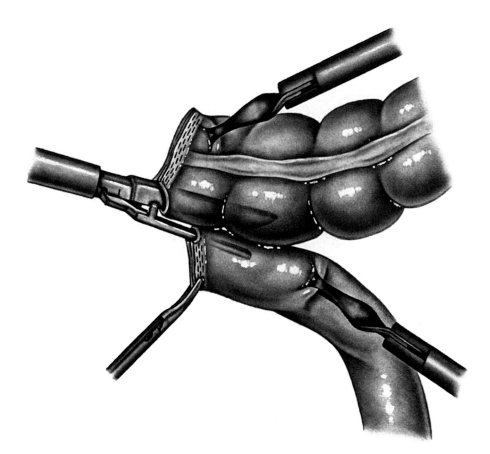

Figure VL-1,F: One fork of the ENDO GIA instrument is inserted into each bowel lumen. The bowel ends are evenly aligned, the antimesenteric bowel walls are compressed, and the instrument is closed and activated, creating the side-to-side anastomosis.

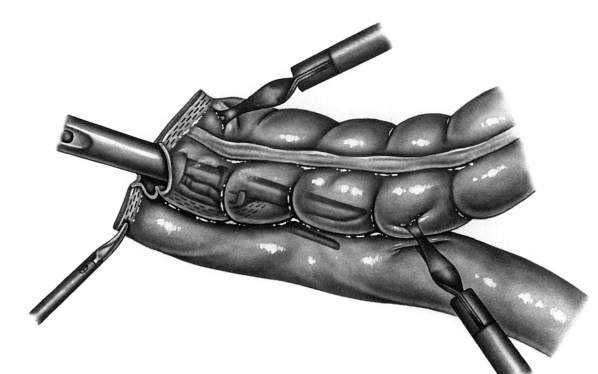

Figure VL-1,G: If needed, the ENDO GIA instrument is reloaded and reapplied into the crotch of the first side-to-side anastomosis, to achieve the desired anastomotic size. Care is taken to continue in the same antimesenteric plane.

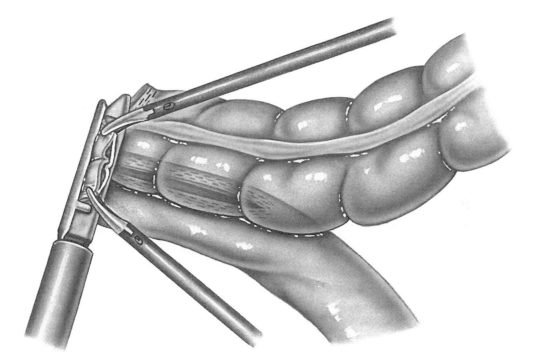

Figure VL-1,H: Following removal of the instrument, the anastomosis is inspected to ensure hemostasis. The common opening is closed with a final application of the ENDO GIA instrument ensuring inclusion of the entire circumference of the remaining opening, all tissue layers, and the ends of the linear, anastomosing staple lines. This functional end-to-end procedure lends itself well to dissimilar bowel lumina. Following inspection of the abdomen, the specimen is removed through a port site, enlarged for this purpose, and this abdominal wall incision and all trocar sites are closed.

RIGHT HEMICOLECTOMY AND "ANASTOMOSE-RESECTION INTEGREE" ANASTOMOSIS FIRST - RESECTION SECOND

As the transfer of laparoscopic techniques from excision of an organ only, such as in cholecystectomy, to excision and reconstruction, such as in bowel resection and anastomosis, took place, one timid step at a time, it soon became obvious that maintaining abdominal wall integrity, except for trocar sites, using the technique shown in Figure VL-1, A to H, was not possible in most instances. The main reason was the evacuation of the specimen, which required an, euphemistically called, enlargement of a trocar site, when in fact this became a true abdominal incision. And, if such an incision was necessary, why not place it where it would do the most good and use it to accomplish the anastomosis first? This in turn brought the various functional end-to-end anastomoses to the forefront, especially the "anastomose-resection integree," to where it had first been described by Ravitch after right hemicolectomy (Figure V-4, A to F). From this relatively easy first step, this reasoning and its implementation were progressively extended all around the colon, to the peritoneal reflexion. At that level, anastomoses could be accomplished with the circular anastomosing instrument, placed through the rectum or through an anterior perineal approach. The new advance was called laparoscopically assisted, as opposed to laparoscopically guided, operation such as excision only in cholecystectomy for instance. While this did represent a compromise, it did not infringe significantly on the advantages and benefits that we have come to expect from the minimally invasive approach.

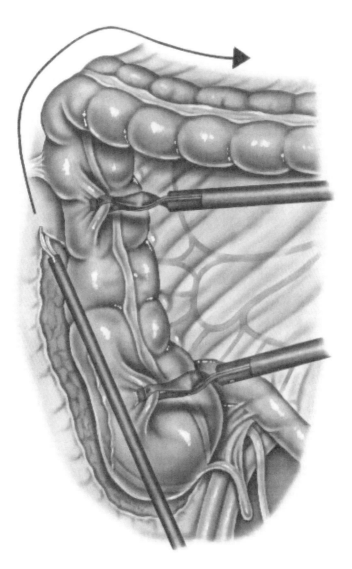

Figure VL-2,A: The laparoscopically assisted mode is especially suited to procedures where a "no-touch" technique with early ligation of vessels is not of the essence, as in this operation for cecal villous adenoma. The parietal peritoneum is incised along the cecum and ascending colon and the bowel to be resected is freed by sharp dissection and blunt elevation from the retroperitoneal area. This mobilization extends from the cecal cul-de-sac, around the hepatic flexure to the point of planned transection of the transverse colon. The mesenteric vessels are doubly ligated on the patient side and singularly ligated on the specimen side as in Figure VL-1,C, or are severed together with the mesocolon by serial applications of the ENDO GIA instrument. The bowel is kept in continuity.

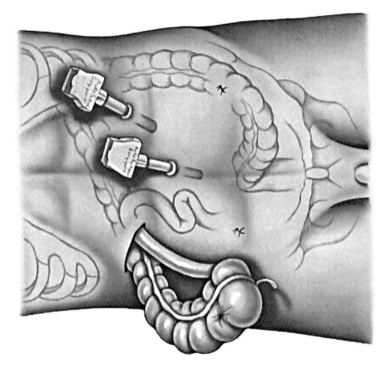

Figure VL-2,B: The port site in the right upper quadrant (RUQ) is enlarged in the direction that opens the abdominal wall to the shortest route for the future specimen to be brought out onto the abdominal wall without tension, usually an oblique, right lateral incision. The loop of terminal ileum, cecum, and ascending colon is drawn out of the abdominal cavity, until the demarcation between viable and non-viable bowel reaches above skin level.

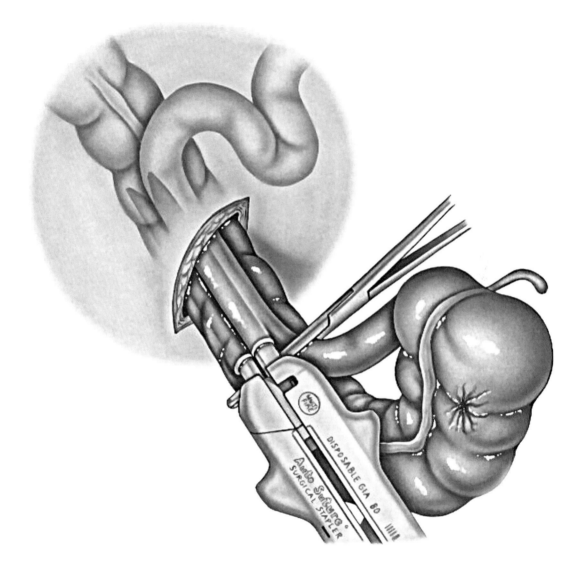

Figure VL-2,C: Both limbs of the ileocecal-right colon loop are positioned so that the viable afferent and efferent bowel segments reach the same horizontal level and can be held in place with a clamp across both bowel walls at the transition point between viable and nonviable bowel. A stab wound is placed through the antimesenteric-mesocolic walls of both viable bowel limbs, central to the occluding clamp, and one fork of the GIA instrument is inserted into each lumen. For this anastomosis to be accomplished, the forks of the instrument can advance inside the bowel lumina, beyond skin level into the abdominal cavity, taking care to have the GIA forks hug the antimesenteric-mesocolic crest of the respective bowel limbs. The instrument is closed and activated, creating the side-to-side anastomosis.

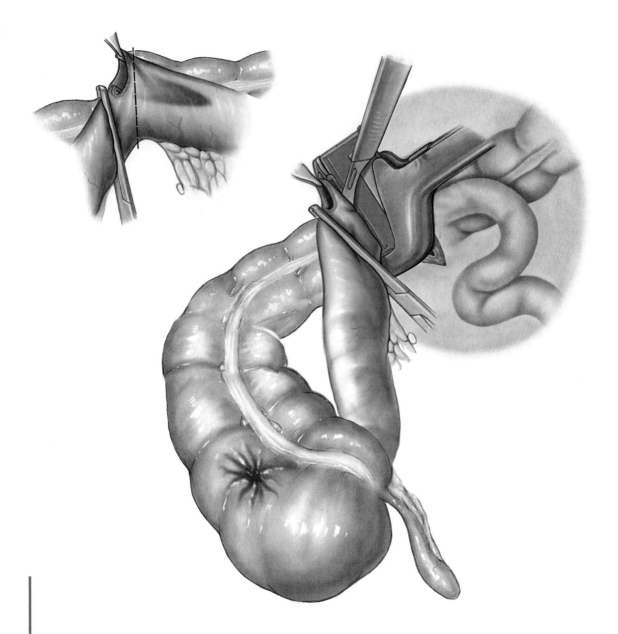

Figure VL-2,D: A V-shaped anastomosis can now be fashioned by placing traction on the anterior anastomotic staple line and holding the anastomotic lines in opposition. The anastomosis is inspected for hemostasis.

Figure VL-2,E: The anastomosis is completed by closure with a linear stapler and resection of the specimen along the edge of the stapler.

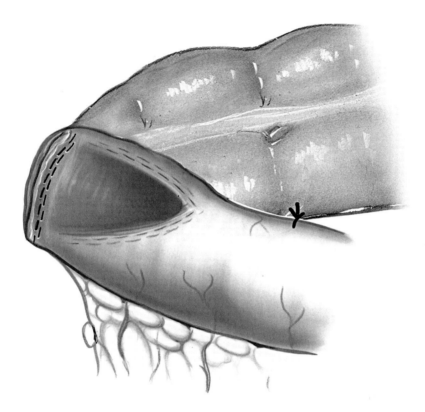

Figure VL-2,F: The result is a relatively wide V-shaped anastomotic cross-section. To avoid tension at the base of the anastomosis, we have routinely placed a manual suture to reinforce the angle formed by the afferent with the efferent bowel loop. In the early days, this was jokingly referred to as "Steichen's good night stitch," because when he knew that this suture had been placed, Steichen slept well that night. The bowel is repositioned into the abdomen and the RUQ incision is closed. The pneumoperitoneum is recreated, and the abdomen is inspected before removal of the remaining ports and closure of their access sites.

EXCISION OF MECKEL'S DIVERTICULUM

As in a laparotomy, many a Meckel's diverticulum will be discovered as an incidental finding during laparoscopy. In the era of mechanical sutures, its removal then and now represents no more than a footnote to the main operative procedure, undertaken for a different indication at the time. Acute Meckel's diverticulitis often mimics appendicitis and would be discovered either during laparoscopy done for reasons of differential diagnosis, or during the routine "running" of the terminal ileum through a McBurney incision that had uncovered a normal appendix. Acute bleeding from a Meckel's diverticulum would usually have led to a "work-up" for "GI bleeding of unknown origin," which in today's world could conceivably be accelerated by a triple endoscopy; negative esophagogastroscopy, colonoscopy negative for the source of bleeding, positive laparoscopy. In the patient with bleeding from a Meckel's diverticulum, a segmental small bowel resection is indicated to remove all of the ectopic gastric mucosa.

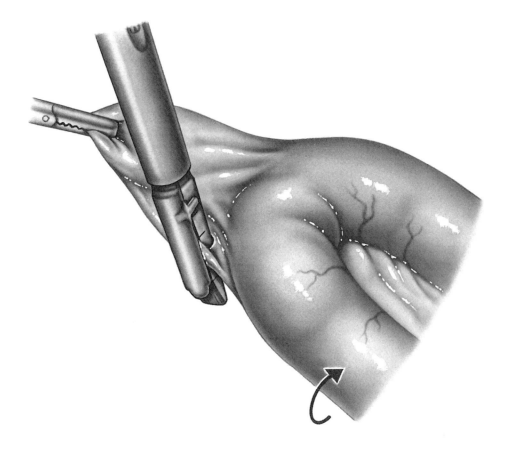

Figure VL-3,A: Through a window made in the avascular area of the diverticular mesentery, the vessels to the diverticulum are caught between the open forks of the ENDO GIA stapler. The instrument is closed and activated, ligating and dividing the diverticular vessels.

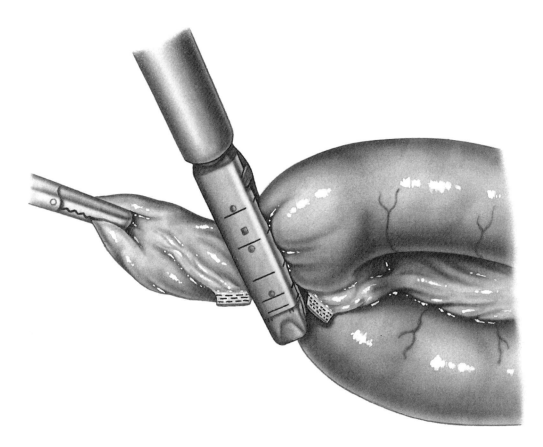

Figure VL-3,B: The base of the diverticulum is closed transversely and simultaneously transected with a second application of the ENDO GIA instrument, placed at a right angle to the long axis of the small bowel. Another application of the ENDO GIA stapler may be necessary with a wide-mouthed diverticulum.

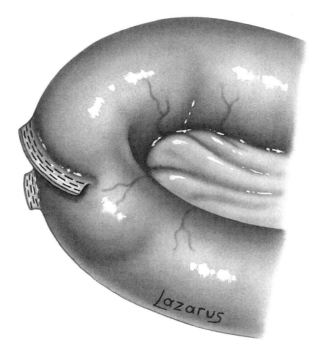

Figure VL-3,C: The transverse closure reduces any potential for stricture formation. If there has been unexplained GI bleeding in the patient's past, a segmental small bowel resection with functional end-to-end anastomosis is indicated.

APPENDECTOMY

As noted in the introduction to this chapter, the unbridled enthusiasm for laparoscopic appendectomy was somewhat reigned in by the realization that in boys, adolescents, and even grown men, the treatment of appendicitis, strongly suspected clinically and confirmed by sonography, is served just as well by an old fashioned McBurney incision as by the elaborate doings of laparoscopy. However, the slightest concern with a differential diagnosis, such as ruptured duodenal ulcer, Meckel's diverticulitis, or adnexal disease in women, does return laparoscopy to its full importance, both as a diagnostic and therapeutic measure.

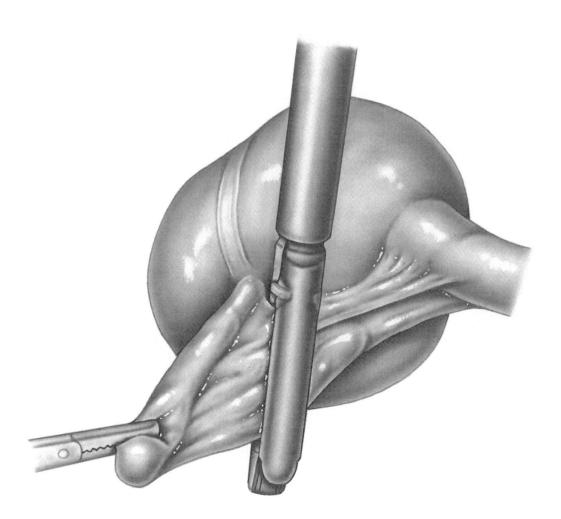

Figure VL-4,A: A window is established through the avascular space of the mesoappendix and the appendiceal vessels with surrounding fat are caught between the open forks of the ENDO GIA stapler. The instrument is closed and activated, ligating and dividing the appendiceal vessels.

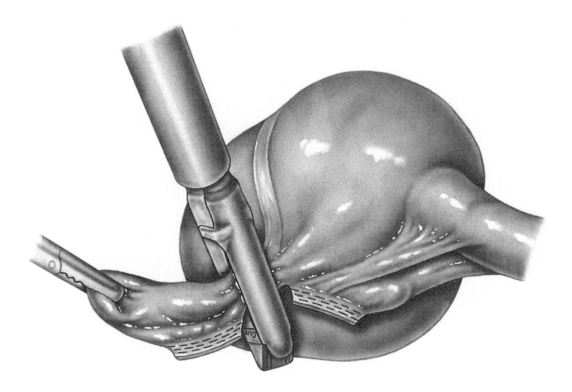

Figure VL-4,B: The appendicocecal junction is closed and transected with a second application of the ENDO GIA instrument.

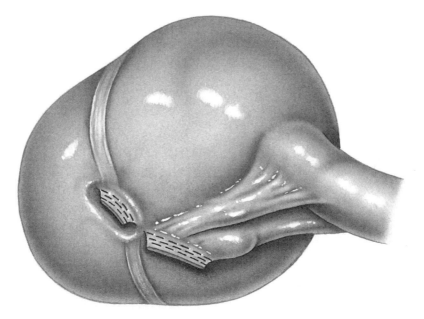

Figure VL-4,C: The closures are inspected for hemostasis, and the abdomen is inspected prior to removal of the ports and closure of the abdominal port sites.

TRANSVERSE COLECTOMY

Transverse colectomy by laparoscopic techniques with a functional end-to-end ileo-distal transverse colostomy, or even ileo-upper descending colostomy, is usually more involved during the dissection phase because of the redundancy of the transverse colon and the weight of the attached greater omentum. The anastomosis would be the mirror image of the ileo-proximal transverse colostomy (Figure VL-2,C), done through an incision in the left upper quadrant.

LEFT COLECTOMY

The "anastomose-resection integree" possesses enough versatility to facilitate any length of colon resection, from a segmental resection to a full hemicolectomy, as long as sufficient mobile bowel remains at either end of the resection to accommodate the forks of the GIA instrument chosen, without putting undue tension on the "crotch" or base of the side-to-side anastomosis. However, if a left colectomy of any extent requires the mobilization of the upper or mid-rectum for the sole purpose of accommodating one fork of the GIA instrument, then the use of the circular anastomosing instrument through the anorectum would be preferable.

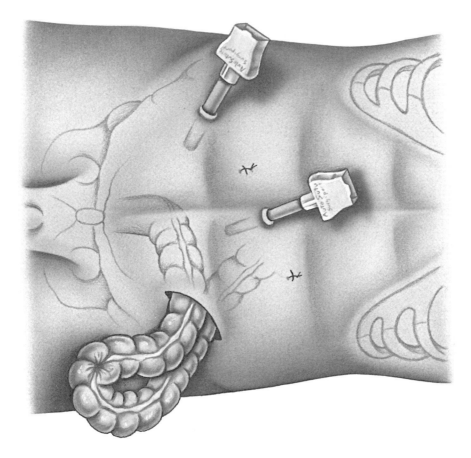

Figure VL-5,A: Following laparoscopic mobilization of the left colon, the port site in the left lower quadrant is enlarged to allow elevation of the specimen to be resected on to the abdominal wall. In case of a left hemicolectomy, this mobilization extends from proximal to mid-colon to the peritoneal reflexion and involves most importantly the liberation of the splenic flexure. Our tolerance is low for leaving the splenic flexure in place, especially if it is very high, for almost any length of left colon resected, regardless of manual or mechanical sutures used in the reconstruction because of our concern for tension on the anastomosis, more so with the GIA stapler than with the EEA instrument.

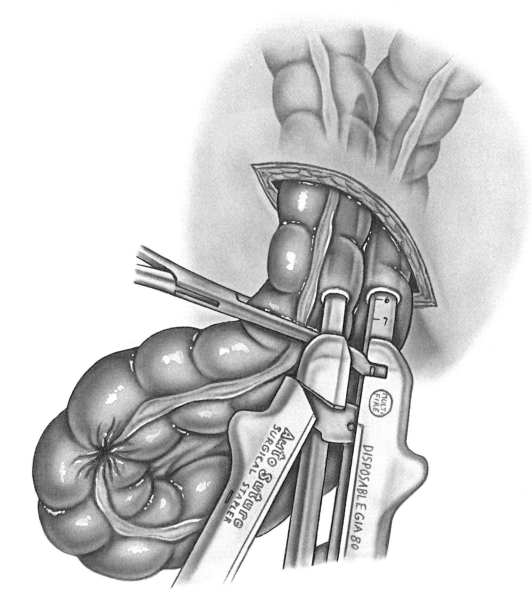

Figure VL-5,B: The specimen loop is arranged so that the projected proximal and distal points of transection are parallel and in direct contact (omega configuration) and held in place with a clamp across both bowel walls, at the respective levels of demarcation between viable and non viable bowel. A stab wound is placed into the antimesenteric border of each bowel wall on the viable side of the holding clamp and one fork of the GIA instrument is inserted into each lumen. The instrument is closed and activated, creating the side-to-side anastomosis. Following removal of the GIA instrument, the anastomosis is inspected for hemostasis and integrity.

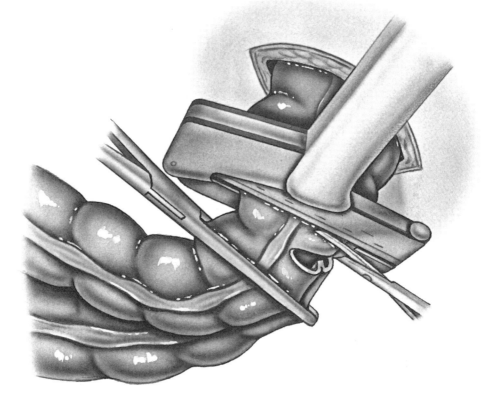

Figure VL-5,C: A V-shaped anastomosis is obtained by pulling the ends of the anastomotic staple lines in opposite directions at the GIA introduction site. The anastomosis is completed by closing this introduction site transversely with a linear stapler and resecting the specimen using the instrument edge as a cutting guide.

Figure VL-5,D: The resulting anastomosis is examined prior to reposition into the abdomen and closure of the incision.

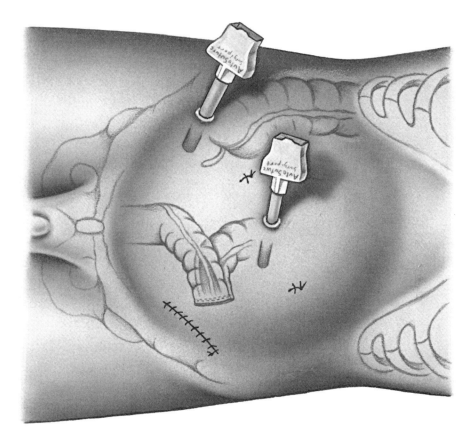

Figure VL-5,E: The pneumoperitoneum is recreated, and the abdomen is inspected prior to removal of the remaining ports and closure of their access sites.

SIGMOID COLON RESECTION

Resection of the sigmoid colon only is most often possible and indicated for benign conditions, diverticulitis being the most frequent culprit. If resection becomes necessary during the acute stage of inflammation or severe bleeding from diverticula, a Hartmann procedure or a vented colocolostomy (Figure V-17, A to H) may be the better part of valor. Reconstruction of continuity after a Hartmann procedure is shown in Figure VL-8, A to K, and of a vented colocolostomy in Figure V-17, J to L. If repeated episodes of diverticulitis or one incidence of hemorrhage from diverticula can be calmed down non-operatively without endangering the patient's survival, then the elective laparoscopic approach has moved the decision making in favor of earlier operative treatment because of its lesser intraoperative stress and earlier return to full activity for the patient.

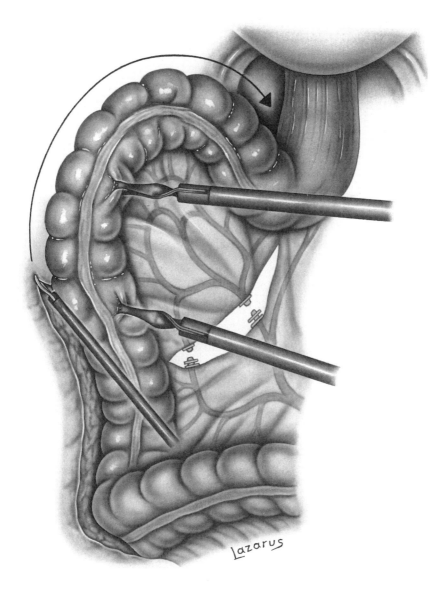

Figure VL-6,A: The sigmoid and superior rectal vessels are ligated high at their origins from the inferior mesenteric artery and vein, and the parietal peritoneum along the sigmoid colon is incised. Depending on the amount of proximal bowel to be removed, the inferior mesenteric artery and vein can be ligated just distal to the ascending left colic artery.

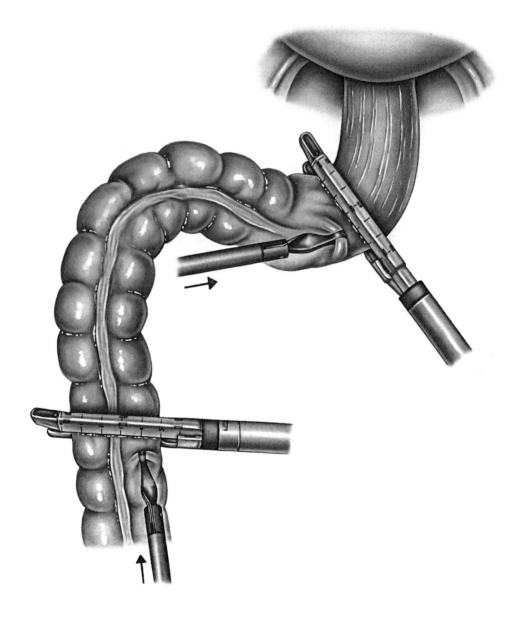

Figure VL-6,B: After liberating the splenic flexure, the descending colon to sigmoid junction and the upper rectum are closed with applications of the ENDO TA instrument at each site, as determined by the separation of viable from non-viable bowel. The rectum is irrigated to clear it of debris.

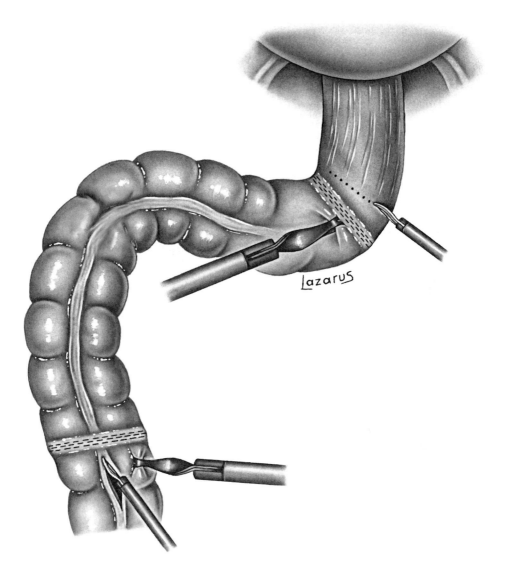

Figure VL-6,C: A longitudinal colotomy is made in the descending colon just proximal to the staple line closure separating the descending from the sigmoid colon. The rectum is incised transversely just below the staple line across the rectum.

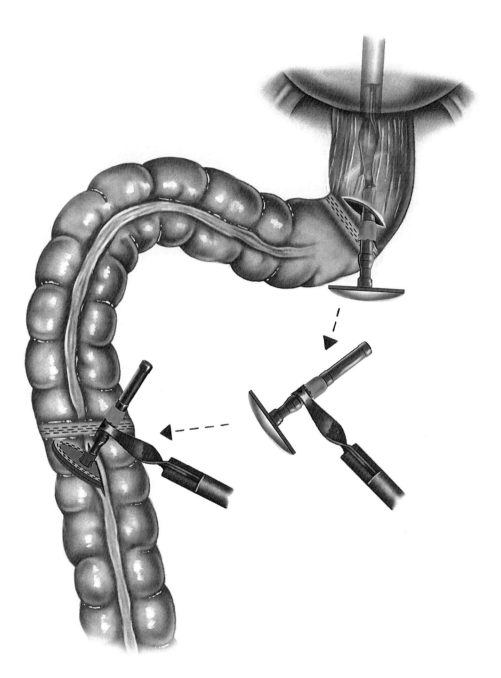

Figure VL-6,D: The anvil of the EEA instrument is introduced transanally and recovered through the rectal incision with a Babcock clamp, inside the abdominal cavity. It is then "portaged" toward the longitudinal opening in the descending colon and is introduced into the proximal bowel lumen with its shaft pointing downward.

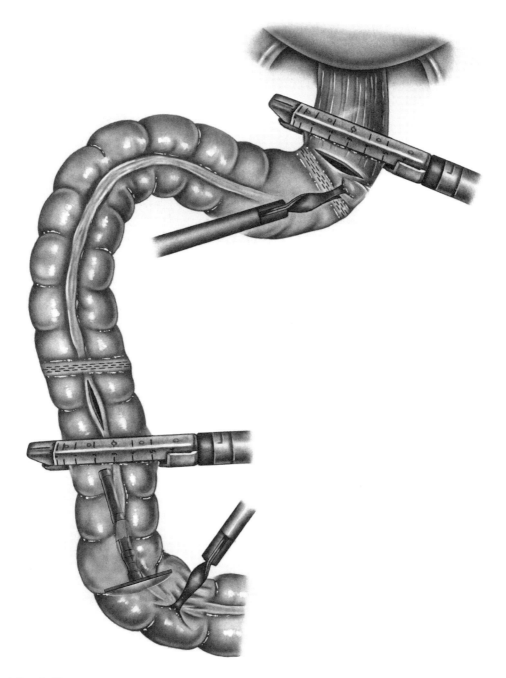

Figure VL-6,E: The ENDO GIA instrument is used to close and transect the descending colon and the upper rectum, above the longitudinal colotomy and below the transverse proctotomy respectively. A partially closed Babcock clamp placed above the anvil on the proximal colon will prevent it from drifting upward. The specimen is removed through a McBurney incision.

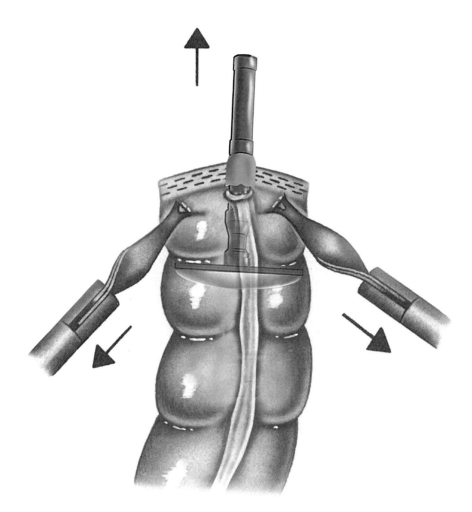

Figure VL-6,F: The incision is closed and the pneumoperitoneum recreated. A small stab wound just above the midpoint of the proximal staple line that closes the descending colon facilitates the grasping of the anvil shaft, which is then drawn through the incision. The anvil is firmly positioned against the staple line, inside the bowel.

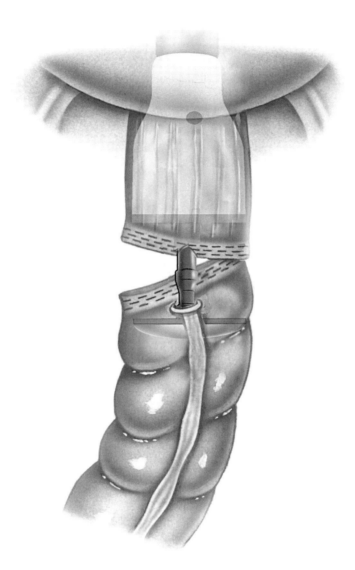

Figure VL-6,G: The EEA instrument carrying the cartridge only is introduced transanally to the level of the closed rectum. The center rod of the instrument is advanced to perforate the rectum at the midpoint of the staple line. The anvil shaft is grasped with a Babcock clamp, advanced into the pelvis with the proximal colon trailing behind, and mated to the center rod of the EEA instrument. The anvil and cartridge are closed against each other, making sure that the colon and rectal staple lines overlap in a cross or X-like fashion. The instrument is activated, placing the double circles of anastomosing staples and cutting the stoma across the linear staple lines; triple stapling anastomosis (Figure V-22, A to F). The anvil and cartridge are separated, and the instrument is removed. The anastomosis is inspected for hemostasis by transanal endoscopy and the "donuts" of tissue containing staples removed with the instrument are inspected for circular continuity and inclusion of all tissue layers.

ANTERIOR RECTOSIGMOID RESECTION

In the early development period of operative laparoscopy, anterior rectosigmoid resection was given a wide berth, especially if a low anastomosis was required after resection for malignancy. However, as experience grew, it soon became obvious that the enhanced and enlarged picture of the pelvis obtained with the endoscope provided a better visual exploration, facilitating sharp dissection along the "holy plane" of Heald, recognizing and respecting the nervi erigentes and ureters, while accomplishing a total mesorectal excision. The "no-touch" technique with early ligation of the inferior mesenteric vessels at the aorta or preferably just distal to the origin of the ascending left colic artery, appeared to be easier than in open procedures. The double and triple stapling techniques with the EEA, GIA, and TA instruments, as well as the anterior perineal approach for low and mid-rectal lesions, lent great comfort and assurance to the execution of an operation that is always fraught with technical hurdles, regardless of an open or closed approach. The evacuation of the specimen can take place through the rectum or an abdominal incision for benign disease, or through an anterior perineotomy for malignant disease. This approach maintains the integrity of the abdominal wall, except for the trocar sites, and is much better tolerated by the patient than an abdominal incision.

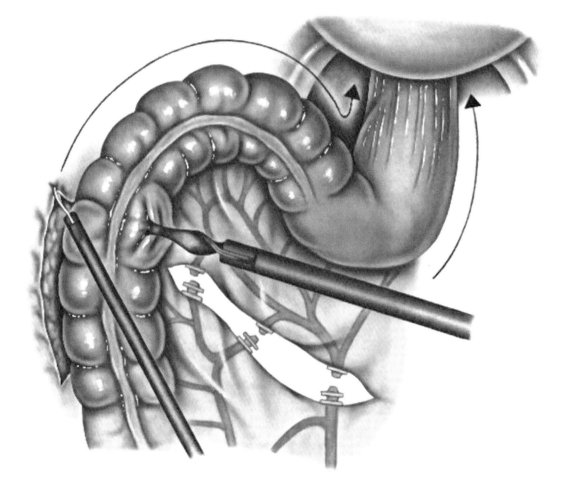

Figure VL-7,A: The inferior mesenteric vessels are individually ligated at the level of the aorta, or preferably just distal to the origin of the ascending left colic artery. The parietal peritoneum is incised along the sigmoid colon and on both sides of the rectum, by staying wide of the bowel if the operation is for malignancy. In order to accomplish an anastomosis deep in the pelvis, free of tension, the splenic flexure is liberated. The avascular plane between rectum and mesorectum and autonomic nerve plexuses is developed by sharp dissection under endoscopic surveillance. In case of a malignancy, the lower pole of the tumor permitting, a rectal cuff of 2 to 3 cm can be saved above the sphincters.

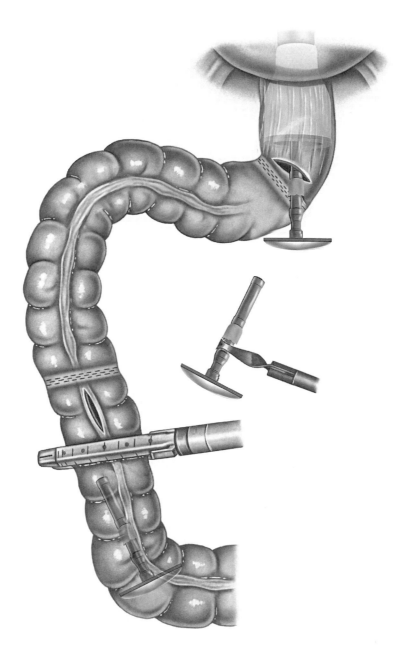

Figure VL-7,B: If the operation is performed for benign disease, but does require anastomosis to the upper or mid-rectum, because of the presence of low-lying diverticula for instance, as in Figure VL-6, B and C, the rectum is closed transversely below the lowest lesion. It is irrigated through the anus to clear it of debris and of cancer cells in case of malignancy. Similarly, the junction of sigmoid and descending colon is closed with the ENDO TA stapler at the demarcation of viable with non-viable tissue. A transverse proctotomy is performed below the rectal closure and a longitudinal colotomy above the colosigmoid closure. The anvil is advanced through the anus, rectum, and proctotomy into the peritoneal cavity, grasped with a Babcock clamp, and transported to the proximal colotomy where it is placed into the bowel lumen with shaft pointing downward.

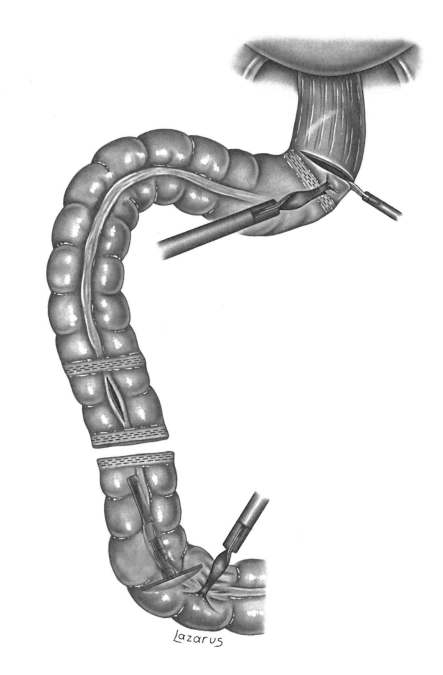

Figure VL-7,C: The descending colon is closed and transected with the ENDO GIA instrument above the longitudinal colotomy, and the anvil is prevented from migrating intraluminally by gently constricting the bowel lumen with a partially closed Babcock clamp above the anvil's location inside the bowel. The anterior transverse proctotomy is transformed into a complete transection of the rectum, by severing its posterior wall at this level.

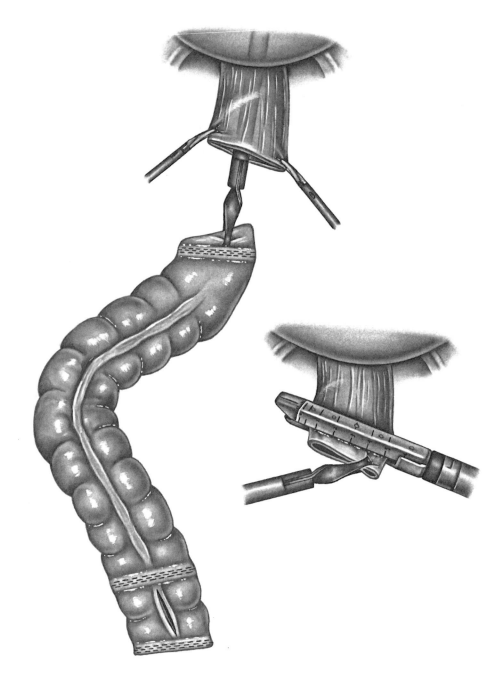

Figure VL-7,D: If this procedure is for benign disease and the specimen is not bulky, it may be retrieved through the rectum and anus, by keeping the open rectum from invaginating as the specimen passes, with two laparoscopic graspers (Azagra, Goergen, 1990). If the specimen is bulky, it is removed through a McBurney incision. If an incision was necessary, it is closed and the pneumoperitoneum is re-established.

Figure VL-7,E: The rectum is closed definitively with the ENDO GIA instrument, and the resected free border is removed.

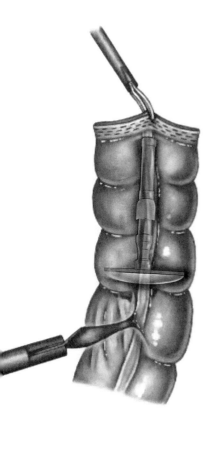

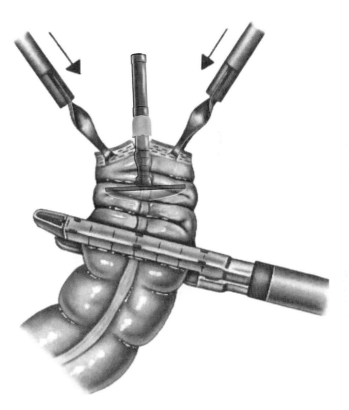

Figure VL-7,F,G: The center of the proximal staple line on the descending colon is incised, and the shaft of the anvil is coaxed through the resulting opening by telescoping the bowel end over the shaft and using a spent ENDO GIA stapler as a stop across the bowel, to prevent the anvil from being pushed superiorly. The anvil shaft is grasped and pulled gently to position the anvil against the staple line inside the proximal colon.

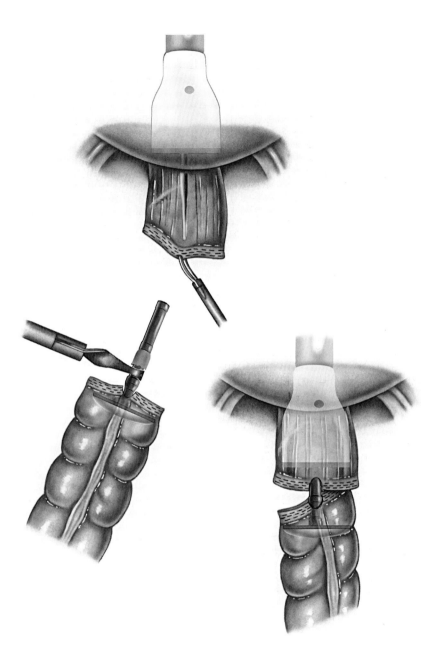

Figure VL-7,H: The EEA instrument, without anvil, is placed through the anus into the rectum and the center rod is advanced through an incision in the center of the rectal linear closure. The anvil, carrying the proximal colon, is transported into the pelvis by a gentle pull on its shaft with a Babcock clamp.

Figure VL-7,I: The anvil shaft and center rod are mated and the instrument is closed, taking care to arrange the linear staple lines in colon and rectum in an X-like mode.

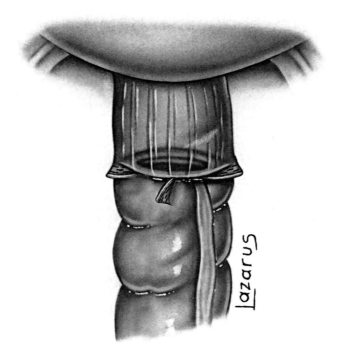

Figure VL-7,J: The tissue on the instrument is checked and the instrument is activated, opened, and removed. The final result is a patent, circular anastomosis achieved with the triple stapling technique that can be easily inspected by anorectal endoscopy.

If the anterior resection is done for a malignancy, especially of the mid or lower rectum, then the access to the safe margin below the tumor is impossible under the constraints of a laparoscopic approach only, in spite of all the sophisticated articulating instruments that are presently available. A conversion to an open operation after the dissection and mobilization of all the rectosigmoid has taken place laparoscopically, would negate all the benefits acquired up to that point by the more tedious laparoscopic procedure and would question the wisdom of a primary minimally invasive approach, when in fact the low position of the tumor could have warned of the obstacles and technical challenges that lay ahead. A nice solution to elude the stifling of operative progress between a hard tumor and the pelvic wall is the anterior perineal approach to reconstruction of bowel continuity of Welter.

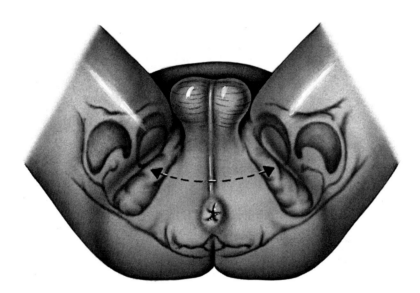

Figure VL-7,K: The transverse perineal skin incision is placed 2 cm in front of the anus. The anobulbar raphe is divided, exposing the anterior aspect of the rectum, just above the external sphincter and deep to the levator ani muscles on both sides. It is then easy to pass an index finger and thumb by blunt dissection around the rectum, through the surrounding areolar tissue, by staying more distal on the dorsal side, in order to free the circumference of the rectum in the same plane. A rubber drain is passed around the rectum to facilitate exposure. In the elderly patient, the weakened levator ani muscles can usually be retracted, whereas in younger patients, strong and fleshy levator muscles may require partial section on one side and later repair, to provide enough exposure. At this point in the procedure, the decision is made as to the feasibility of anterior resection and coloanal anastomosis, versus abdominoperineal amputation, by measuring the safe margin beyond the tumor. It should at least be 3 cm between the lower pole of the tumor and the upper aspect of the striated sphincter, to favor anastomosis over amputation.

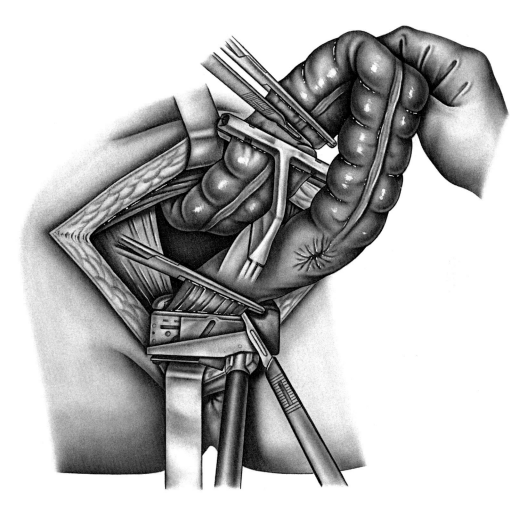

Figure VL-7,L: With gentle traction on the rubber drain around the rectum, the abdominal dissection is completed from below. The tumor-bearing rectum is pulled down, through the perineotomy, until healthy sigmoid colon with intact blood supply reaches the upper aperture of the perineum. At this stage of the procedure, the rectum is closed with the Roticulator instrument above the striated sphincter (the range can be up to 2 cm above this level, depending on the safe margin to the tumor). The rectum is transected between the distal linear stapler, using its upper edge as a guide, and a proximal Kocher clamp, occluding the rectum on the specimen side. The Roticulator instrument is removed. On the sigmoid side of the specimen, the Purstring instrument is placed onto healthy viable bowel and is activated to place a purse-string suture. The sigmoid is then transected between the Purstring instrument and a clamp, occluding the specimen, this time by using the lower edge of the Purstring instrument as a guide.

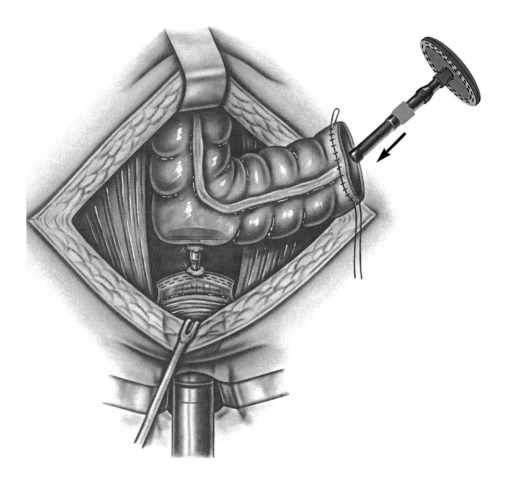

Figure VL-7,M: The specimen is removed and examined one more time for adequate tumor margins. If they are unsatisfactory, the abdominoperineal amputation is completed by resecting the anus with the usual perineal collar of skin, subcutaneous fat, and muscles. If satisfactory, the anastomosis is now started by placing the EEA instrument, with cartridge only, into the anal canal and advancing the center rod through or near the center of the linear staple line. The anvil is passed into the open proximal sigmoid colon and secured with its shaft leading and brought out through a stab wound in the antimesosigmoid wall of the normally curved sigmoid that approaches the anal closure with the least tension.

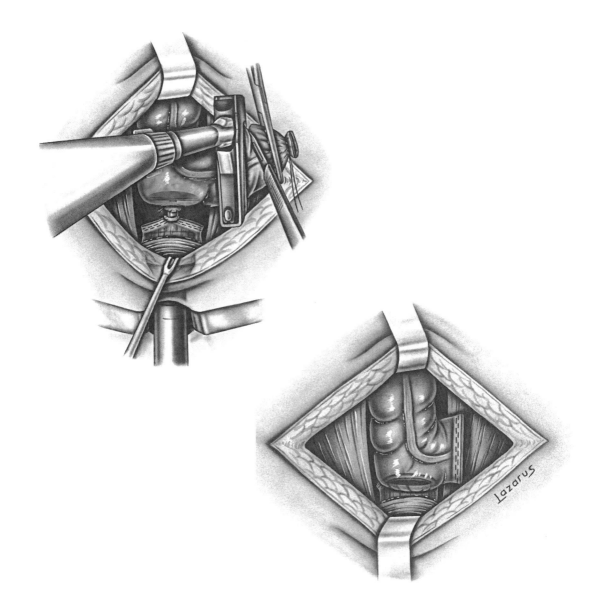

Figure VL-7,N: The EEA anvil and cartridge are closed against each other and the instrument is activated, producing a side-to-end sigmoid-anal anastomosis. The excess sigmoid colon beyond the lateral border of this anastomosis is closed with the linear stapler and excised using the Roticulator or TA stapler used as a guide for the scalpel.

Figure VL-7,O: The final result is a side-to-end sigmoid-anal anastomosis, following the pattern of the double stapling technique. By using the antimesocolic side of the normally curved sigmoid colon, the diameter of the bowel adds some 3 to 4 cm to the length of the afferent limb and relieves some of the tension that often complicates such extremely low anastomoses.

HARTMANN PROCEDURE
SECOND-STAGE REPAIR OF BOWEL CONTINUITY

The operation described by Henri Hartmann of Paris nearly a hundred years ago and consisting of resection of the offending colosigmoid colon, proximal end colostomy, and distal mucus fistula or closure of the rectal stump has extricated surgeon and patient alike from many a delicate, often unexpected emergency situation. Even with today's advances in anastomotic technique, reconstruction of bowel continuity is always tedious, at times treacherous, because of adhesions around the colostomy and especially around the rectal stump, if it was closed and allowed to retract deep into the pelvis. Whenever possible, we have therefore preferred the construction of a vented colocolostomy (Figure V-17, A to H) at the primary procedure, which could then be easily closed under local anesthesia in a second stage (Figure V-17, J to L). However, the vented colocolostomy requires the presence of sufficient colon at both ends to accommodate the arms of the longer GIA instrument; sometimes a remote aim under emergency conditions. Therefore, the Hartmann procedure remains a proven and safe solution to many an urgent operative impasse.

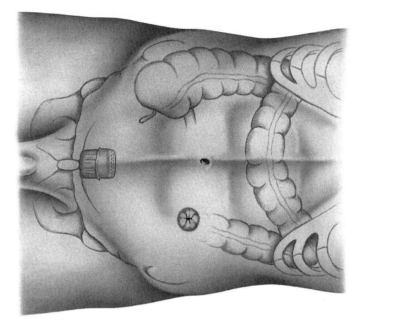

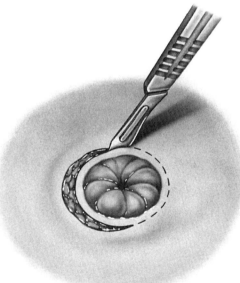

Figure VL-8,A,B: The end colostomy is freed up by a circular incision around the mucocutaneous rim. This incision is deepened through subcutaneous, fascial, muscular, and peritoneal layers into the abdominal cavity, by staying close to the bowel.

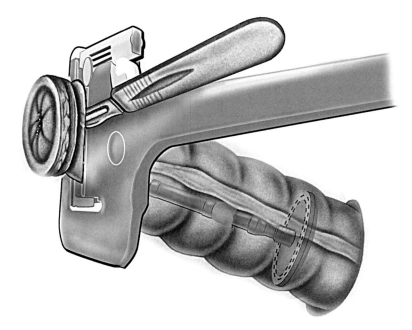

Figure VL-8,C: As and when the descending colon has become sufficiently mobile, the anvil with its shaft is placed into the open lumen of the descending colon. The colon is closed transversely in soft bowel, below the crown of the scarred and fibrous mucocutaneous ostium.

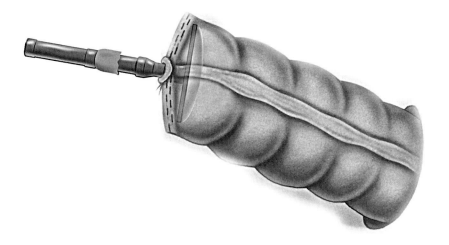

Figure VL-8,D: The shaft of the anvil is pulled out through an incision in the center of the linear colon closure and secured with a purse-string suture. The bowel end, together with the anvil, is repositioned into the left lateral abdominal gutter.

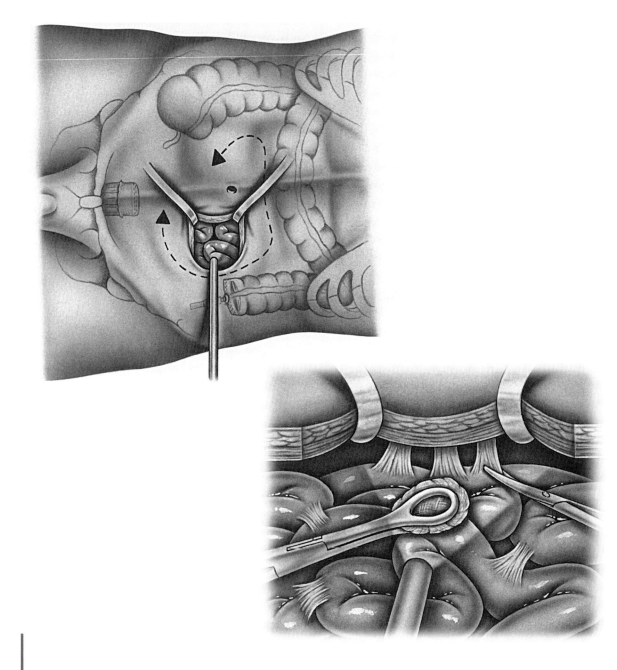

Figure VL-8,E,F: With the circular opening of the previous colostomy site retracted and elevated, the pericolostomy adhesions are severed under direct vision. As this lysis of adhesions progresses medially and inferiorly, the laparoscope, held through the circular opening, can help in advancing the lysis significantly.

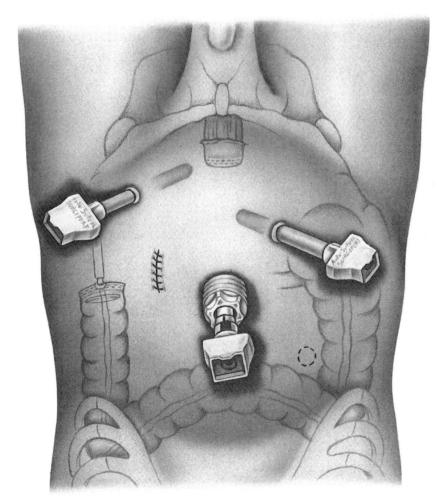

Figure VL-8,G: After the adhesions to the anterior abdominal wall have been severed sufficiently, the laparoscope through the umbilicus and two trocars, one in the right lower quadrant and the other in the left suprapubic area, are placed, under direct vision through the circular opening. The opening is then closed and pneumoperitoneum is established.

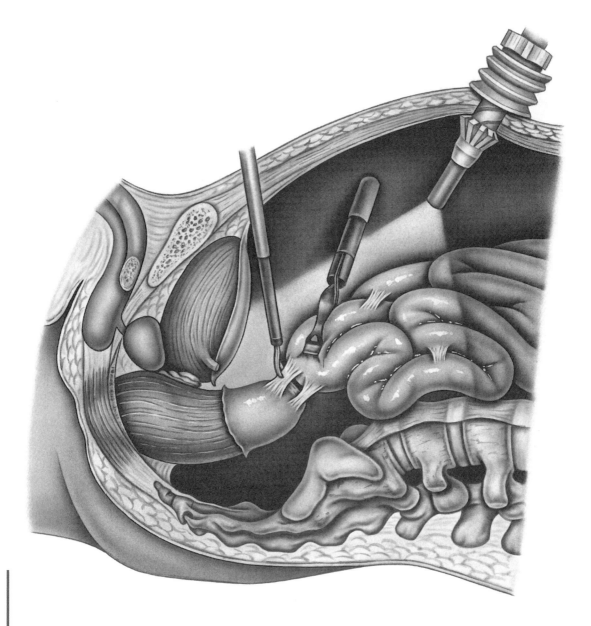

Figure VL-8,H: Under laparoscopic vision, the rectal stump is freed from surrounding adhesions. This part of the procedure may be helped by introducing the EEA instrument with cartridge into the rectum and moving it left to right and anteriorly.

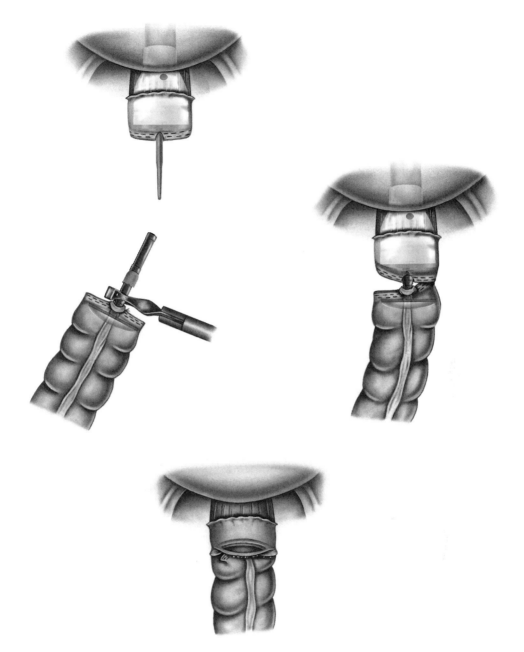

Figure VL-8,I,J,K: After the stump has been cleared sufficiently, the center rod is advanced through the scar at the stump cul-de-sac and the proximal colon with anvil and extruding shaft is brought into the pelvis. The center rod and anvil shaft are mated, the instrument closed, taking care to arrange the linear staple lines in colon and rectum in an X-like mode, and activated performing a triple stapling anastomosis.

SUGGESTED READING

Adloff M, Arnaud JP, Beehary S, Turbelin JM. Side-to-end anastomosis in low anterior resection with the EEA stapler. *Dis Colon Rectum.* 1980;23:456-458.

Beart RW, Kelly KA. Randomized prospective evaluation of the EEA stapler for colorectal anastomosis. *Am J Surg.* 1981;141:143-147

Fazio VW. Advances in the surgery of rectal carcinoma utilizing the surgical stapler. In: Spratt JS, ed. *Neoplasms of the Colon, Rectum and Anus.* Philadelphia, Pa: WB Saunders; 1984:268-288.

Fuchs KH, Engemann R, Thiede A. *Klammernahttechnik in der Chirugie.* Berlin, Heidelberg: Springer Verlag; 1993.

Goldberg SM, Gordon PH, Nivatvongs S. *Essentials of Ano-rectal Surgery.* Philadelphia, Pa: J B Lippincott; 1980.

Goligher JC. Use of circular stapling gun with perianal insertion of ano-rectal pursestring suture for construction of low colorectal or colo-anal anastomoses. *Br J Surg.* 1979;66:501-504.

Heald RJ, Husband EM, Ryall RDH. The mesorectum in rectal cancer surgery: the clue to pelvic recurrence. *Br J Surg.* 1982;69:13-16.

Heald, RJ. The holy plane of rectal surgery. *JR Soc Med.* 1988;81:503-508.

Knight CD, Griffen FD. An improved technique for low anterior resection of the rectum using the EEA stapler. *Surgery.* 1980;88:710.

Kremer K, Platzer W, Schreiber HW, Steichen FM, eds. *Minimally Invasive Abdominal Surgery.* Stuttgart-New York: Thieme Verlag; 2001.

Lepor H, Gregerman M, Crosby R, Mostofi RK, Walsh PC. Precise localization of autonomic nerves from pelvic plexus to corpora cavernosa: detailed anatomical study of adult male pelvis. *J Urol.* 1985;133:207-212.

Milsom JW, Bohm B. *Laparoscopic Colorectal Surgery.* New York, Berlin: Springer; 1996

Mouiel J, Katkhouda N, Gugenheim J, Fabiani D, le Goff D, Benizri E, Gouboux B. Subtotal colectomy followed by stapled ceco-rectal anastomosis. In: *Current Practice of Surgical Stapling,* Philadelphia, Pa: Lea & Febiger Publishers; 1991:285-288.

Ravitch MM, Steichen FM, eds. Symposium on surgical stapling techniques. *Surg Clin North Am*. 1984;64:423-621.

Ravitch MM, Steichen FM, eds. *Principles and Practice of Surgical Stapling*. Chicago, Ill: Yearbook Medical Publishers; 1987: chap 20-27.

Ravitch MM, Ong TH, Gazzola L. A new precise and rapid technique of intestinal resection and anastomosis with staples. *Surg Gynecol Obstet*. 1974;139:6-10.

Ravitch MM, Steichen FM, Welter R, eds. *Current Practice of Surgical Stapling*. Philadelphia, London: Lea & Febiger; 1991.

Scott-Conner CEH, ed. *The Sages Manual: Fundamentals of Laparoscopy and GI Endoscopy*. New York, Berlin: Springer; 1999.

Steichen FM. The use of staplers in anatomical side-to-side and functional end-to-end enteroanastomoses. *Surgery*. 1968;64:948-953.

Steichen FM, Loubeau JM, Stremple JF. The continent ileal reservoir of Kock. *Surg Rounds*, 1978 Sept:10-18.

Steichen FM, Ravitch MM. *Stapling in Surgery*. Chicago, Ill: Yearbook Medical Publishers; 1984.

Steichen FM, Spigland NA, Nunez D. The modified Duhamel operation for Hirschsprung's disease performed entirely with mechanical sutures. *J Ped Surg*. 1987;22:436.

Steichen FM, Barber HK, Loubeau JM, Iraci JC. Bricker-Johnston sigmoid colon graft for repair of postradiation rectovaginal fistula and stricture performed with mechanical sutures. *Dis Colon Rectum*. 1992;35:599-603.

Steichen FM, Welter R, eds. *Minimally Invasive Surgery and New Technology*. St. Louis, Mo: Quality Medical Publishing, Inc.; 1994.

Turbelin JM, Arnaud JP, Welter R, Adloff MA. Etude comparative des surfaces anastomotiques obtenues par utilisation des sutures mecaniques en chirurgie digestive. *J Chir (Paris)*. 1980;117:541-546.

Ulrich B, Winter J. *Klammernahttechnik in Thorax und Abdomen, vol 99 Praktische Chirurgie*. Stuttgart, Enke Verlag; 1986.

Welter R, Patel JC. *Chirurgie Mecanique Digestive*. Paris, Masson; 1985.

INDEX

G

GIA instrument
  for appendectomy, 36-37
  for cecal volvulus correction, 98-101
  for colon closure and transection, 78, 107, 135, 143
  for creation of an ileal J-pouch, 162-164
  description of, 2-3
  for Duhamel operation, 198
  for functional end-to-end. See Functional end-to-end anastomosis
  for ileo-transverse colon bypass operation, 48-50
  for rectovaginal fistula repair, 206
  for side-to-side anastomosis, 28-33, 134-141, 142-146
  for terminal ileum resection and ileocecostomy, 38-43
  for vented colocolostomy, 90-95
  V-shaped closure of introduction site in functional end-to-end anastomosis, 14-21

H

Hartmann procedure, 91, 258-265
Hemicolectomy
  laparoscopic, 210-217, 218-223
  right
      anastomose-resection integree and, 218-223
      ileo-transverse colostomy and, 54-59, 210-217
Hirschsprung's disease, Duhamel operation for, 192-199

I

Ileal J-pouch
  transabdominal construction and ileoanal anastomosis, 162-171,
  transperineal anastomosis to anal canal, 178-183
  transperineal anastomosis to anus, 184-189
Ileoanal anastomosis, 162-171, 178-183, 184-189
Ileocecal anastomotic fistula, 44-47
Ileocecostomy, 38-43
Ileocolostomy, 48-53, 54-59, 210-217, 218-223
Ileostomy, 88-89, 170
Ileotomy, 51-52
Ileo-transverse colon
  anastomosis following resection, 54-59, 210-217, 218-223
  bypass operation, 48-53
Ileum, terminal, 38-43, 44-47, 54-59, 210-217, 218-223
Intestinal bypass, 28-33, 48-53

J

J-pouch, ileal, 162-171, 178-183, 184-189